Eye Tracking Methodology:
Theory and Practice

Springer
London
Berlin
Heidelberg
New York
Hong Kong
Milan
Paris
Tokyo

Andrew T. Duchowski

Eye Tracking Methodology: Theory and Practice

Springer

Andrew T. Duchowski
Department of Computer Science, 451 Edwards Hall, Clemson University,
Clemson, SC 29634-0974, USA

British Library Cataloguing in Publication Data
Duchowski, Andrew T.
  Eye tracking methodology : theory and practice
  1.Human-computer interaction 2.Visual perception 3.Eye -
  Movements 4.Tracking (Psychology)
  I.Title
  004'.019
  ISBN 1852336668
Library of Congress Cataloging-in-Publication Data
Duchowski, Andrew T.
  Eye tracking methodology : theory and practice / Andrew T. Duchowski
    p. cm.
  Includes bibliographical references and index.
  ISBN 1-85233-666-8 (alk. paper)
    1. Human-centred interaction. 2. Computer graphics. 3. Computer vision. 4. Visual
  perception. I. Title.
  QA76.9.H85 D83 2002
  004'.01'9--dc21   2002030645

ISBN 1-85233-666-8 Springer-Verlag London Berlin Heidelberg
a member of BertelsmannSpringer Science+Business Media GmbH
http://www.springer.co.uk

Typesetting: Camera ready by author
Printed and bound at the Athenæum Press Ltd., Gateshead, Tyne and Wear
34/3830-543210 Printed on acid-free paper SPIN 10883214

Preface

The scope of the book falls within a fairly narrow Human-Computer Interaction domain (i.e., describing a particular input modality), however, it spans a broad range of inter-disciplinary research and application topics. There are at least three domains that stand to benefit from eye tracking research: visual perception, human-computer interaction, and computer graphics. The amalgamation of these topics forms a symbiotic relationship. Graphical techniques provide a means of generating rich sets of visual stimuli ranging from 2D imagery to 3D immersive virtual worlds while research exploring visual attention and perception in turn influences the generation of artificial scenes and worlds. Applications derived from these disciplines create a powerful Human-Computer Interaction modality, namely interaction based on knowledge of the user's gaze.

Recent advancements in eye tracking technology, specifically the availability of cheaper, faster, more accurate and easier to use trackers, have inspired increased eye movement and eye tracking research efforts. However, although eye trackers offer a uniquely objective view of overt human visual and attentional processes, eye trackers have not yet gained widespread use beyond work conducted at various research laboratories. This lack of acceptance is due in part to two reasons: first, the use of an eye tracker in an applied experimental setting is not a widely taught subject. Hence, there is a need for a book that may help in providing training. It is not uncommon for enthusiastic purchasers of eye tracking equipment to become discouraged with their newly bought equipment when they find it difficult to set up and operate. Only a few academic departments (e.g., Psychology, Computer Science) offer any kind of instruction in the use of eye tracking devices. Second, to exacerbate the lack of training in eye tracking methodology, even fewer sources of instruction exist for system development. Setting up an eye tracking lab and integrating the eye tracker into an available computer system for development of *gaze-contingent* applications is a fairly complicated endeavor, similar to the development and integration of Virtual Reality programs. Thus far, it appears no textbook other than this one exists providing this type of low-level information.

The goal of this book is to provide technical details for implementation of a gaze-contingent system, couched in the theoretical context of eye movements, visual perception, and visual attention. The text started out as the author's personal notes on the integration of a commercial eye tracker into a Virtual Reality graphics system. These technical considerations comprise the middle chapters of the book and include details of integrating a commercial eye tracker into both a 3D Virtual Environment, and a 2D image display application. The surrounding theoretical review chapters grew from notes developed for an interdisciplinary Eye Tracking Methodology course offered to both undergraduates and graduates from four disciplines: Psychology, Marketing, Industrial Engineering, and Computer Science. An early form of these notes was presented as a short course at the Association for Computing Machinery (ACM) Special Interest Group on Graphics' SIGGRAPH conference, 23-28 July 2000, New Orleans, LA.

Overview

The book is divided into three parts, presented thematically in a top-down fashion, providing first an Introduction to the Human Visual System (Part I), then briefly surveying Eye Tracking Systems (Part II), and finally ending by summarizing a number of Eye Tracking Applications (Part III).

In the first part, Introduction to the Human Visual System (HVS), the book covers the concept of visual attention, mainly from a historical perspective. The first chapter focuses on the dichotomy of foveal and peripheral vision (the "what" vs. the "where"). While this chapter covers easily observable attentional phenomena, the next chapter covers the neurological substrate of the HVS presenting the low-level neurological elements implicated in dynamic human vision. This chapter discusses the primary dual pathways, the parvo- and magno-cellular channels, which loosely correspond to the flow of visual information permitted by the retinal fovea and periphery. Following this description of the visual "hardware", observable characteristics of human vision are summarized in the following chapter on visual perception. Here, results obtained mainly from psychophysics are summarized, distinguishing foveal and peripheral visual perception. The first part ends by discussing the mechanism responsible for shifting the fovea, namely eye movements. Having established the neurological and psychophysical context for eye movements, the following chapter on the taxonomy and models of eye movements gives the common terms for the most basic of eye movements along with a signal-analytic description of recordable eye movement waveforms.

The second part of the book, Eye Tracking Systems, presents a brief survey of the main types of available eye tracking devices, followed by a detailed technical description of the requirements for system installation and application program development. These details are mainly applicable to video-based, corneal-reflection eye trackers, the most widely available and most affordable type of eye trackers. This part of the book offers information for the development of two general systems: one for binocular 3D eye tracking in Virtual Reality, and one for monocular 2D eye tracking over a 2D display (e.g., a television monitor on which graphical information can be displayed). This part of the book ends with a description of system calibration, data collection, and analysis.

The third part of the book surveys a number of interesting and challenging eye tracking applications. Applications identified in this part are drawn from Psychology, Human Factors, Marketing and Advertising, Human-Computer Interaction and Collaborative Systems, and Computer Graphics and Virtual Reality.

How to Read this Book

The intended audience for this book is an inter-disciplinary one, aimed particularly at those interested in Psychology, Marketing, Industrial Engineering, and Computer Science. Indeed, this text is meant for undergraduates and graduates from these disciplines enrolled in a course dealing with eye tracking, such as the Eye Tracking Methodology course developed by the author at Clemson University. In this course, typically all chapters are covered, but not necessarily in the order presented in the text. In such a course, the order of chapters may be as follows.

First, Part III is presented outlining various eye tracking applications. Normally, this part should give the reader motivation for design and implementation of a one-semester eye tracking project. Coverage of this part of the book is usually supplanted by readings of research papers from various sources. For example, papers may be selected from the following conferences: The Computer Graphics Proceedings, the proceedings of the annual Association for Computing Machinery (ACM) Special Interest Group on Graphics and Interactive Techniques (SIGGRAPH) conference series, the proceedings of the ACM Special Interest Group on Human-Computer Interaction (SIGCHI), the proceedings of the Human Factors and Ergonomics Society, and the Eye Tracking Research & Applications (ETRA) conference.

To speed up development of a gaze-contingent application, Part II follows the presentation of Part III, dealing in the technical details of eye tracker application development. The types of applications that can be expected of students will depend mainly on the programming expertise represented by members of inter-disciplinary student teams. For example, in the Eye Tracking Methodology course at Clemson, teams are formed by joining Computer Science students with one or more of the other representatives enrolled in the class, i.e., from Marketing, Psychology, or Industrial Engineering. While all group members decide on a project, students studying the latter subjects are mainly responsible for the design and analysis of the eventual eye tracking experiment.

Once implementation of an eye tracking application has commenced, Part I of the text is covered, giving students the necessary theoretical context for the eye tracking pilot study. Thus, although the book is arranged "top-down", the course proceeds "bottom-up".

The book is also suitable for researchers interested in setting up an eye tracking laboratory and/or using eye trackers for conducting experiments. Since members with these goals may also come from diverse disciplines such as Marketing, Psychology, Industrial Engineering, and Computer Science, not all parts of the book may be suitable for all readers. More technically oriented readers will want to pay particular attention to the middle sections of the book which detail system installation and implementation of eye tracking application software. Readers not directly involved with such low-level details may wish to omit these sections and concentrate more on the theoretical and historical aspects given in the front sections of the book. The latter part of the book, dealing with eye tracking applications, should be suitable for all readers since it presents examples of current eye tracking research.

Acknowledgments

This work was supported in part by a University Innovation grant (# 1-20-1906-51-4087), NASA Ames task (# NCC 2-1114), and NSF CAREER award # 9984278.

The preparation of this book has been assisted by many people, including Keith Karn, Roel Vertegaal, Dorion Liston, and Keith Rayner who provided comments on early editions of the text-in-progress. Later versions of the draft were reviewed by external reviewers to whom I express my gratitude for their comments greatly improved the final version of the text. Special thanks go to David Wooding for his careful and thorough review of the text.

I would like to thank the team at Springer for helping me compose the text. Thanks go to Beverly Ford and Karen Borthwick for egging me on to compose the text and to Rosie Kemp and Melanie Jackson for helping me with the final stages of publication.

Special thanks go to Bruce McCormick, who always emphasized the importance of writing during my doctoral studies at Texas A&M University, College Station, TX. Finally, special thanks go to Corey, my wife, for patiently listening to my various ramblings on eye movements, and for being an extremely patient eye tracking subject :)

I have gained considerable pleasure and enjoyment in putting the information I've gathered and learned on paper. I hope that readers of this text derive similar pleasure in exploring vision and eye movements as I have, and they go on to implementing ever interesting and fascinating projects—have fun!

Clemson, SC, June 2002   *Andrew T. Duchowski*

Contents

List of Figures

List of Tables

Part I

Introduction to the Human Visual System (HVS)

1. Visual Attention

In approaching the topic of eye tracking, we first have to consider the motivation for recording human eye movements. That is, why is eye tracking important? Simply put, we move our eyes to bring a particular portion of the visible field of view into high resolution so that we may see in fine detail whatever is at the central direction of gaze. Most often we also divert our attention to that point so that we can focus our concentration (if only for a very brief moment) on the object or region of interest. Thus, we may presume that if we can track someone's eye movements, we can follow along the path of attention deployed by the observer. This may give us some insight into what the observer found interesting, i.e., what drew their attention, and perhaps even provide a clue as to how that person perceived whatever scene s/he was viewing.

By examining attention and the neural mechanisms involved in visual attention, the first two chapters of this book present motivation for the study of eye movements from two perspectives: (1) a psychological viewpoint examining attentional behavior and its history of study (presented briefly in this chapter); (2) a physiological perspective on the neural mechanisms responsible for driving attentional behavior (covered in the next chapter). In sum, both introductory chapters establish the psychological and physiological basis for the movements of the eyes.

To begin formulating an understanding of an observer's attentional processes, it is instructive to first establish a rudimentary or at least intuitive sense of what attention is, and whether or not the movement of the eyes does in fact disclose anything about the inner cognitive process known as visual attention.

Visual attention has been studied for over a hundred years. A good qualitative definition of visual attention was given by the psychologist James:

> Everyone knows what attention is. It is the taking possession by the mind, in clear and vivid form, of one out of what seem several simultaneously possible objects or trains of thought. Focalization, concen-

tration, of consciousness are of its essence. It implies withdrawal from some things in order to deal effectively with others...

When the things are apprehended by the *senses*, the number of them that can be attended to at once is small, '*Pluribus intentus, minor est ad singula sensus.*'

—James (1981)

The Latin phrase used above by James roughly translates to "*Many filtered into few for perception*". The faculty implied as the filter is attention.

Humans are finite beings that cannot attend to all things at once. In general, attention is used to focus our mental capacities on selections of the sensory input so that the mind can successfully process the stimulus of interest. Our capacity for information processing is limited. The brain processes sensory input by concentrating on specific components of the entire sensory realm so that interesting sights, sounds, smells, etc., may be examined with greater attention to detail than peripheral stimuli. This is particularly true of vision. Visual scene inspection is performed *minutatim*, not *in toto*. That is, human vision is a piecemeal process relying on the perceptual integration of small regions to construct a coherent representation of the whole.

In this chapter, attention is recounted from a historical perspective following the narrative found in Heijden (1992). The discussion focuses on attentional mechanisms involved in vision, with emphasis on two main components of visual attention, namely the "what" and the "where".

1.1 Visual Attention: A Historical Review

The phenomenon of visual attention has been studied for over a century. Early studies of attention were technologically limited to simple ocular observations and oftentimes to introspection. Since then the field has grown to an interdisciplinary subject involving the disciplines of psychophysics, cognitive neuroscience, and computer science, to name three. This section presents a qualitative historical background of visual attention.

1.1.1 Von Helmholtz's "where"

At the second half of the 19th century, Von Helmholtz (1925) posited visual attention as an essential mechanism of visual perception. In his *Treatise on*

Physiological Optics, he notes, "We let our eyes roam continually over the visual field, because that is the only way we can see as distinctly as possible all the individual parts of the field in turn." Noting that attention is concerned with a small region of space, Von Helmholtz observed visual attention's natural tendency to wander to new things. He also remarked that attention can be controlled by a conscious and voluntary effort, allowing attention to peripheral objects without making eye movements to that object. Von Helmholtz was mainly concerned with eye movements to spatial locations, or the "where" of visual attention. In essence, although visual attention can be consciously directed to peripheral objects, eye movements reflect the will to inspect these objects in fine detail. In this sense, eye movements provide evidence of overt visual attention.

1.1.2 James' "what"

In contrast to Von Helmholtz's ideas, James (1981) believed attention to be a more internally covert mechanism akin to imagination, anticipation, or in general, thought. James defined attention mainly in terms of the "what", or the identity, meaning, or expectation associated with the focus of attention. James favored the active and voluntary aspects of attention although he also recognized its passive, reflexive, non-voluntary and effortless qualities.

Both views of attention, which are not mutually exclusive, bear significantly on contemporary concepts of visual attention. The "what" and "where" of attention roughly correspond to foveal (James) and parafoveal (Von Helmholtz) aspects of visual attention, respectively. This dichotomous view of vision is particularly relevant to a bottom-up or feature-driven explanation of visual attention. That is, when considering an image stimulus, we may consider certain regions in the image which will attract one's attention. These regions may initially be perceived parafoveally, in a sense requesting further detailed inspection through foveal vision. In this sense, peripherally located image features may drive attention in terms of "where" to look next, so that we may identify "what" detail is present at those locations.

The dual "what" and "where" feature-driven view of vision is a useful preliminary metaphor for visual attention, and indeed it has formed the basis for creating computational models of visual attention, which typically simulate so-called low-level, or bottom-up visual characteristics. However, this view of attention is rather simplistic. It must be stressed that a complete model of visual attention involves high-level visual and cognitive functions. That is, visual attention cannot simply be explained through the sole consideration of visual features. There are higher-level *intentional* factors involved (e.g., related to possibly voluntary, pre-conceived cognitive factors that drive attention).

1.1.3 Gibson's "how"

In the 1940s Gibson (1941) proposed a third factor of visual attention centered on intention. Gibson's proposition dealt with a viewer's advance preparation as to whether to react and if so, how, and with what class of responses. This component of attention explained the ability to vary the intention to react while keeping the expectation of the stimulus object fixed, and conversely, the ability to vary the expectation of the stimulus object while keeping the intention to react fixed. Experiments involving ambiguous stimuli typically evoke these reactions. For example, if the viewer is made to expect words describing animals, then the misprint "sael" will be read as "seal". Changing the expectation to words describing ships or boats invokes the perception of "sail". The reactive nature of Gibson's variant of attention specifies the "what to do", or "how" to react behavior based on the viewer's preconceptions or attitude. This variant of visual attention is particularly relevant to the design of experiments. On the one hand it is important to consider the viewer's perceptual expectation of the stimulus, as (possibly) influenced by the experimenter's instructions.

1.1.4 Broadbent's "selective filter"

Attention, in one sense, is seen as a "selective filter" responsible for regulating sensory information to sensory channels of limited capacity. In the 1950s, Broadbent (1958) performed auditory experiments designed to demonstrate the selective nature of auditory attention. The experiments presented a listener with information arriving simultaneously from two different channels, e.g., the spoken numerals $\{7,2,3\}$ to the left ear, $\{9,4,5\}$ to the right. Broadbent reported listeners' reproductions of either $\{7,2,3,9,4,5\}$, or $\{9,4,5,7,2,3\}$, with no interwoven (alternating channel) responses. Broadbent concluded that information enters in parallel but is then selectively filtered to sensory channels.

1.1.5 Deutsch and Deutsch's "importance weightings"

In contrast to the notion of a selective filter, Deutsch and Deutsch (1963) proposed that all sensory messages are perceptually analyzed at the highest level, precluding a need for a selective filter. Deutsch and Deutsch rejected the selective filter and limited capacity system theory of attention; they reasoned that the filter would need to be at least as complex as the limited capacity system itself. Instead, they proposed the existence of central structures with preset "importance weightings" which determined selection. Deutsch and Deutsch argued that it is not attention as such but the weightings of importance that have a causal role in attention. That is, attentional effects are a result of importance, or relevance, interacting with the information.

It is interesting to note that Broadbent's selective filter generally corresponds to Von Helmholtz's "where", while Deutsch and Deutsch's importance weightings correspond to James' expectation, or the "what". These seemingly opposing ideas were incorporated into a unified theory of attention by Anne Treisman in the 1960s (although not fully recognized until 1971). Treisman brought together the attentional models of Broadbent and Deutsch and Deutsch by specifying two components of attention: the attenuation filter followed by later (central) structures referred to as 'dictionary units'. The attenuation filter is similar to Broadbent's selective filter in that its function is selection of sensory messages. Unlike the selective filter, it does not completely block unwanted messages, but only attenuates them. The later stage dictionary units then process weakened and unweakened messages. These units contain variable thresholds tuned to importance, relevance, and context. Treisman thus brought together the complementary models of attentional unit or selective filter (the "where"), and expectation (the "what").

Up to this point, even though Treisman provided a convincing theory of visual attention, a key problem remained, referred to as the scene integration problem. The scene integration problem poses the following question: even though we may view the visual scene through something like a selective filter, which is limited in its scope, how is it that we can piece together in our minds a fairly coherent scene of the entire visual field? For example, when looking at a group of people in a room such as classroom or at a party, even though it is impossible to gain a detailed view of everyone's face at the same time, nevertheless it is possible to assemble a mental picture of where people are located. Our brains are capable of putting together this mental picture even though the selective filter of vision prevents us from physically doing so in one glance. Another well-known example of the scene integration problem is the Kanizsa (1976) illusion, exemplified in Figure 1.1, named after the person who invented it. Inspecting Figure 1.1, you will see the edges of a triangle, even though the triangle is defined only by the notches in the disks. How this triangle is integrated by the brain is not yet fully understood. That is, while it is known that the scene is inspected piecemeal as evidenced by the movement of the eyes, it is not clear how the "big picture" is assembled, or integrated, in the mind. This is the crux of the scene integration problem. In one view, offered by the Gestalt psychologists, it is hypothesized that recognition of the entire scene is performed by a parallel, one-step process. To examine this hypothesis, a visualization of how a person views such an image (or any other) is particularly helpful. This is the motivation for recording and visualizing a viewer's eye movements. Even though the investigation of eye movements dates back to 1907 (Dodge, 1907),[1] a clear depiction of eye movements would not be available until 1967 (see Chapter 5 for a survey of eye tracking techniques). This early eye movement visualization, discussed below, shows the importance of eye movement recording not only for its expressive power of depicting one's visual scanning characteristics, but also for its influence on theories of visual attention and perception.

Fig. 1.1. The Kanizsa illusion.

1.1.6 Yarbus and Noton and Stark's "scanpaths"

Early diagrammatic depictions of recorded eye movements helped cast doubt on the Gestalt hypothesis that recognition is a parallel, one-step process. The Gestalt view of recognition is a holistic one suggesting that vision relies to a great extent on the tendency to group objects. Although well known visual illusions exist to support this view (e.g., subjective contours of the Kanizsa figure, see Figure 1.1 above and Kanizsa (1976)), early eye movement recordings showed that visual recognition is at least partially serial in nature.

---

[1] As cited in Gregory (1990).

Yarbus (1967) measured subjects' eye movements over an image after giving subjects specific questions related to the image. Such a picture is shown in Figure 1.2. Questions posed to subjects included a range of queries specific to the situation, e.g., are the people in the image related, what are they wearing, what will they have to eat, etc. The eye movements Yarbus recorded demonstrated sequential viewing patterns over particular regions in the image.

Noton and Stark (1971a, 1971b) performed their own eye movement measurements over images and coined the observed patterns "scanpaths". Their work extended Yarbus' results by showing that even without leading questions subjects tend to fixate identifiable regions of interest, or "informative details". Furthermore, scanpaths showed that the order of eye movements over these regions is quite variable. That is, given a picture of a square, subjects will fixate on the corners, although the order in which the corners are viewed differs from viewer to viewer and even differs between consecutive observations made by the same individual.

In contrast to the Gestalt view, Yarbus' and Noton and Stark's work suggests that a coherent picture of the visual field is constructed piecemeal through the assembly of serially viewed regions of interest. Noton and Stark's results support James' "what" of visual attention. With respect to eye movements, the "what" corresponds to regions of interest selectively filtered by foveal vision for detailed processing.

1.1.7 Posner's "spotlight"

Contrary to the serial "what" of visual attention, the orienting, or the "where", is performed in parallel (Posner, Snyder, & Davidson, 1980). Posner et al suggested an attentional mechanism able to move about the scene in a manner similar to a "spotlight". The spotlight, being limited in its spatial extent, seems to fit well with Noton and Stark's empirical identification of foveal regions of interest. Posner et al, however, dissociate the spotlight from foveal vision and consider the spotlight an attentional mechanism independent of eye movements. Posner et al identified two aspects of visual attention: the orienting and the detecting of attention. Orienting may be an entirely central (covert, or mental) aspect of attention, while detecting is context-sensitive, requiring contact between the attentional beam and the input signal. The orienting of attention is not always dependent on the movement of the eyes, i.e., it is possible to attend to an object while maintaining gaze elsewhere. According to Posner et al, orientation of attention must be done in parallel and must precede detection.

In each of the traces, the subject was asked to: Trace 1, examine the picture at will; Trace 2, estimate the economic level of the people; Trace 3, estimate the people's ages; Trace 4, guess what the people were doing before the arrival of the visitor; Trace 5, remember the people's clothing; Trace 6, remember the people's (and objects') position in the room; Trace 7, estimate the time since the guest's last visit.

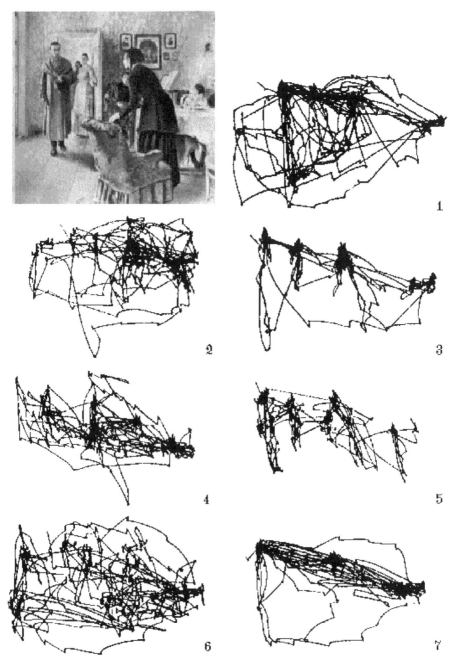

Fig. 1.2. Yarbus' early eye movement recordings. Reprinted from Yarbus (1967) with permission © 1967 Plenum Press.

The dissociation of attention from foveal vision is an important point. In terms of the "what" and the "where", it seems likely that the "what" relates to serial foveal vision. The "where", on the other hand, is a parallel process performed parafoveally, or peripherally, which dictates the next focus of attention.

1.1.8 Treisman's "glue"

Posner et al and Noton and Stark advanced the theory of visual attention along similar lines forged by Von Helmholtz and James (and then Broadbent and Deutsch and Deutsch). Treisman once again brought these concepts together in the Feature Integration Theory of visual attention (Treisman & Gelade, 1980; Treisman, 1986). In essence, attention provides the "glue" which integrates the separated features in a particular location so that the conjunction, i.e., the object, is perceived as a unified whole. Attention selects features from a master map of locations showing *where* all the feature boundaries are located, but not *what* those features are. That is, the master map specifies where things are, but not what they are. The feature map also encodes simple and useful properties of the scene such as color, orientation, size, and stereo distance. Feature Integration Theory, or FIT, is a particularly important theory of visual attention and visual search. Eye tracking is often a significant experimental component which is used to test FIT. Feature Integration Theory, treated as an eye tracking application, is discussed in more detail in Chapter 11.

1.1.9 Kosslyn's "window"

Recently, Kosslyn (1994) proposed a refined model of visual attention. Kosslyn describes attention as a selective aspect of perceptual processing, and proposes an attentional "window" responsible for selecting patterns in the "visual buffer". The window is needed since there is more information in the visual buffer than can be passed downstream, and hence the transmission capacity must be selectively allocated. That is, some information can be passed along, but other information must be filtered out. This notion is similar to Broadbent's selective filter and Treisman's attenuation filter. The novelty of the attentional window is its ability to be adjusted incrementally, i.e., the window is scalable. Another interesting distinction of Kosslyn's model is the hypothesis of a redundant stimulus-based attention-shifting subsystem (e.g., a type of context-sensitive spotlight) in mental imagery. Mental imagery involves the formation of mental maps of objects, or of the environment in general. It is defined as "...the mental invention or recreation of an experience that in at least some respects resembles the experience of actually perceiving an object or an event, either in conjunction with, or in the absence of, direct sensory stimulation"

(Finke, 1989). It is interesting to note that the eyes move during sleep (known as Rapid Eye Movement or REM sleep). Whether this is a manifestation of the use of an internal attentional window during sleep is not known.

1.2 Visual Attention and Eye Movements

Considering visual attention in terms of the "what" and "where", we would expect that eye movements work in a way which supports the dual attentive hypothesis. That is, vision might behave in a cyclical process composed of the following steps:

1. Given a stimulus, such as an image, the entire scene is first "seen" mostly in parallel through peripheral vision and thus mostly at low resolution. At this stage, interesting features may "pop out" in the field of view, in a sense engaging or directing attention to their location for further detailed inspection.
2. Attention is thus turned off or disengaged from the foveal location and the eyes are quickly repositioned to the first region which attracted attention.
3. Once the eyes complete their movement, the fovea is now directed at the region of interest, and attention is now engaged to perceive the feature under inspection at high resolution.

This is a bottom-up model or concept of visual attention. If the model is accurate, one would expect to find regions in the brain that correspond in their function to attentional mechanisms. This issue is further investigated in Chapter 2.

The bottom-up model is at least correct in the sense that it can be said to be a component of natural human vision. In fact, the bottom-up model forms a powerful basis for computational models of visual search. Examples of such models are presented later in the text (see Chapter 11). The bottom-up view of visual attention is, however, incomplete. There are several key points that are not addressed. Consider these questions:

1. Assuming it is only the visual stimulus (e.g., image features) that drives attention, exactly what types of features attract attention?
2. If the visual stimulus were solely responsible for attracting attention, would we ever need the capability of making voluntary eye movements?
3. What is the link between attention and eye movements—is attention always associated with the foveally viewed portion of the visual scene?

To gain insight into the first question, we must examine how our physical visual mechanism (our eyes and brain) responds to visual stimulus. To attempt to validate a model of visual attention, we would need to be able justify the model by identifying regions in the brain that are responsible for carrying out the functionality proposed by the model. For example, we would expect to find regions in the brain that engage and disengage attention as well as those responsible for controlling (i.e., programming, initiating, and terminating) the movements of the eyes. Furthermore, there must be regions in the brain that are responsible for responding to and interpreting the visual stimuli that is captured by the eyes. As will be seen in the following chapters, the Human Visual System (HVS) responds strongly to some types of stimuli (e.g., edges), and weakly to others (e.g., homogeneous areas). The following chapters will show that this response can be predicted to a certain extent by examining the physiology of the HVS. In later chapters we will also see that the human visual response can be measured through a branch of Psychology known as psychophysics. That is, through psychophysics, we can fairly well measure the perceptive power of the Human Visual System.

The bottom-up model of visual attention does not adequately offer answers to the second question because it is limited to mostly bottom-up, or feature-driven aspects of attention. The answer to the second question becomes clearer if we consider a more complete picture of attention involving higher-level cognitive functions. That is, a complete theory of visual attention should also involve those cognitive processes which describe our voluntary *intent* to attend to something, e.g., some portion of the scene. This is a key point which was briefly introduced following the summary of Gibson's work, and which is evident in Yarbus' early scanpaths. It is important to re-iterate that Yarbus' work demonstrated scanpaths which differed with observers' expectations, that is, scanpath characteristics such as their order of progression can be *task-dependent*. People will view a picture differently based on what they are looking for. A complete model or theory of visual attention is beyond the scope of this book, but see Chapter 11 for further insight into theories of visual search, and also for examples of the application of eye trackers to study this question.

Considering the third question opens up a classical problem in eye tracking studies. Because attention is composed of both low-level and high-level functions (one can loosely think of involuntary and voluntary attention, respectively), as Posner and others have observed, humans can voluntarily dissociate attention from the foveal direction of gaze. In fact, astronomers do this regularly to detect faint constellations with the naked eye by looking "off the fovea". Because the periphery is much more sensitive to dim stimulus, faint stars are much more easily seen out of the "corner" of one's eye than when they are viewed centrally. Thus the high-level component of vision may be thought of as a covert component, or a component which is not easily detectable by external observation. This is a well-known problem for eye tracking researchers. An eye tracker can only track the overt movements of the eyes, however, it cannot track the covert movement of visual attention. Thus, in all eye tracking work, a tacit but very important assumption is usually accepted: we assume that attention is linked to foveal gaze direction, but we acknowledge that it may not always be so.

1.3 Summary and Further Reading

An historical account of attention is a prerequisite to forming an intuitive impression of the selective nature of perception. For an excellent historical account of selective visual attention, see Heijden (1992). An earlier and very readable introduction to visual processes is a small paperback by Gregory (1990). For a more neurophysiological perspective, see Kosslyn (1994). Another good text describing early attentional vision is Papathomas, Chubb, Gorea, and Kowler (1995).

The singular idioms describing the selective nature of attention are the "what" and the "where". The "where" of visual attention corresponds to the visual selection of specific regions of interest from the entire visual field for detailed inspection. Notably, this selection is often carried out through the aid of peripheral vision. The "what" of visual attention corresponds to the detailed inspection of the spatial region through a perceptual channel limited in spatial extent. The attentional "what" and "where" duality is relevant to eye tracking studies since scanpaths show the temporal progression of the observer's foveal direction of gaze and therefore depict the observer's instantaneous overt localization of visual attention.

From investigation of visual search, the consensus view is that a parallel, preattentive stage acknowledges the presence of four basic features: color, size, orientation, and presence and/or direction of motion and that features likely to attract attention include edges, corners, but not plain surfaces (see Chapter 11). There is some doubt however, whether human visual search can be described as an integration of independently processed features (Van Orden & DiVita, 1993). Van Orden and DiVita suggest that "...any theory on visual attention must address the fundamental properties of early visual mechanisms." To attempt to quantify the visual system's processing capacity, the neural substrate of the human visual system is examined in the following chapter which surveys the relevant neurological literature.

2. Neurological Substrate of the HVS

Considerable information may be gleaned from the vast neuroscientific literature regarding the functionality (and limitations) of the Human Visual System
(HVS). It is often possible to qualitatively predict observed psychophysical results by studying the underlying visual "hardware". For example, visual spatial
acuity may be roughly estimated from knowledge of the distribution of retinal
photoreceptors. Other characteristics of human vision may also be estimated
from the neural organization of deeper brain structures.

Neurophysiological and psychophysical literature on the human visual system
suggests the field of view is inspected *minutatim* through brief fixations over
small regions of interest. This allows perception of detail through the fovea.
Central foveal vision subtends 1-5° (visual angle) allowing fine scrutiny of
only a small portion of the entire visual field, e.g., only 3% of the size of a
large (21in) computer monitor (seen at ∼60cm viewing distance). Approximately 90% of viewing time is spent in fixations. When visual attention is directed to a new area, fast eye movements (saccades) reposition the fovea. The
dynamics of visual attention probably evolved in harmony with (or perhaps in
response to) the perceptual limitations imposed by the neurological substrate
of the visual system.

The brain is composed of numerous regions classified by their function (Zeki,
1993). A simplified representation of brain regions is shown in Figure 2.1, with
lobe designations stylized in Figure 2.2. The human visual system is functionally described by the connections between retinal and brain regions, known as
visual pathways. Pathways joining multiple brain areas involved in common
visual functions are referred to as streams. Figure 2.1 highlights regions and
pathways relevant to selective visual attention. For clarity, many connections
are omitted. Of particular importance to dynamic visual perception and eye
movements are the following neural regions, summarized in terms of relevance
to attention:

- SC (Superior Colliculus): involved in programming eye movements and contributes to eye movement target selection for both saccades and smooth pursuits (possibly in concert with the Frontal Eye Fields (FEF) and area Lateral Intra-Parietal (LIP)); also remaps auditory space into visual coordinates (presumably for target foveation); with input of motion signals from area MT (see below), the SC is involved in pursuit target selection as well as saccade target selection.
- Area V1 (primary visual cortex): detection of range of stimuli, e.g., principally orientation selection and possibly to a lesser extent color; cellular blob regions (double-opponent color cells) respond to color variations and project to areas V2 and V4 (Livingstone & Hubel, 1988).
- Areas V2, V3, V3A, V4, MT: form, color, and motion processing.
- Area V5/MT (Middle Temporal) and MST (Middle Superior Temporal): furnish large projections to Pons; hence possibly involved in smooth pursuit movements; involved in motion processing—area MT also projects to the colliculus, providing it with motion signals from the entire visual field.
- Area LIP (Lateral Intra-Parietal): contains receptive fields which are corrected ("reset") before execution of saccadic eye movements.
- PPC (Posterior Parietal Complex): involved in fixations.

Connections made to these areas from area V1 can be generally divided into two streams: the dorsal and ventral streams. Loosely, their functional description can be summarized as:

- Dorsal stream: sensorimotor (motion, location) processing (e.g., the attentional "where").
- Ventral stream: cognitive processing (e.g., the attentional "what").

In general attentional terms, the three main neural regions implicated in eye movement programming and their functions are (Palmer, 1999):

- Posterior Parietal Complex: disengages attention;
- SC: relocates attention;
- Pulvinar: engages, or enhances, attention.

In a very simplified view of the brain, it is possible to identify the neural mechanisms involved in visual attention and responsible for the generation of eye movements. First, by examining the structure of the eye, it becomes clear why only the central or foveal region of vision can be perceived at high resolution. Second, signals from foveal and peripheral regions of the eye's retina can be roughly traced along pathways in the brain showing how the brain may process the visual scene. Third, regions in the brain can be identified which are thought to be involved in moving the eyes so that the scene can be examined piecemeal.

In this simplified view of the brain, one can in a sense obtain a complete picture of an "attentional feedback loop", which creates the attentional cycles of disengaging attention, shifting of attention and (usually) the eyes, re-engaging attention and brain regions for processing the region of interest currently being attended to.

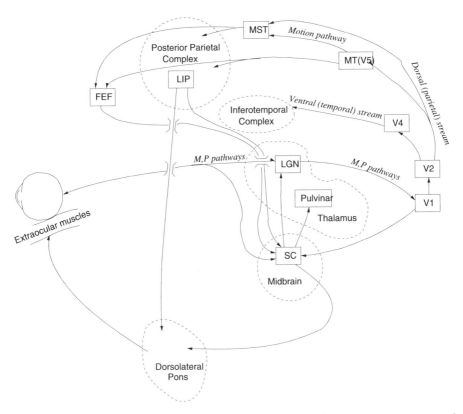

Fig. 2.1. A simplified view of the brain and the visual pathways relevant to eye movements and attention.

The neural substrate of the human visual system is examined in this chapter from the intuitive attentional perspective given above. The human neural "hardware" responsible for visual processing is presented in order roughly following the direction of light and hence information entering the brain. That is, the discussion is presented "front-to-back" starting with a description of the eye and ending with a summary of the visual cortex located at the back of the brain. Emphasis is placed on differentiating the processing capability of foveal and peripheral vision, i.e., the simplified "what" and "where" of visual attention, respectively. However, the reader must be cautioned against

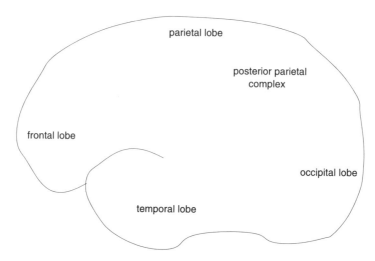

Fig. 2.2. Stylized classification of cortical lobes.

underestimating the complexity of the visual system as presented in this text. The apparent "what" and "where" dual pathways are most probably not independent functional channels. There is a good deal of interconnection and "cross-talk" between these and other related visual centers which deems the dichotomous analysis overly simplistic. Nevertheless, there is a great deal of valuable information to be found in the neurological literature as human vision is undoubtedly the most studied human sense.

2.1 The Eye

Often called "the world's worst camera", the eye, shown in Figure 2.3, suffers from numerous optical imperfections, e.g.,

- spherical aberrations: prismatic effect of peripheral parts of the lens;
- chromatic aberrations: shorter wavelengths (blue) refracted more than longer wavelengths (red);
- curvature of field: a planar object gives rise to a curved image.

However, the eye is also endowed with various mechanisms which reduce degradive effects, e.g.,

- to reduce spherical aberration, the iris acts as a stop, limiting peripheral entry of light rays;
- to overcome chromatic aberration, the eye is typically focused to produce sharp images of intermediate wavelengths;

- to match the effects of curvature of field, the retina is curved compensating for this effect.

The eye is schematically shown in Figure 2.3.

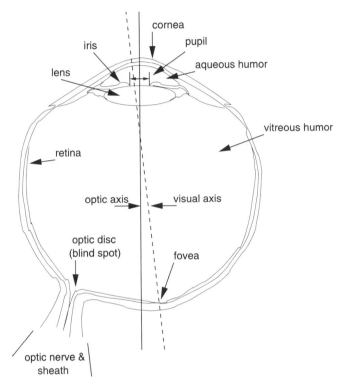

Fig. 2.3. The eye. Adapted from VISUAL PERCEPTION, *1st edition*, by Cornsweet (1970) © 1970. Reprinted with permission of Wadsworth, a division of Thomson Learning: <www.thomsonrights.com>. Fax 800 730-2215.

2.2 The Retina

At the rear interior surface of the eye, the retina contains receptors sensitive to light (photoreceptors) which constitute the first stage of visual perception. Photoreceptors can effectively be thought of as "transducers" converting light energy to electrical impulses (neural signals). Neural signals originating at these receptors lead to deeper visual centers in the brain. Photoreceptors are functionally classified into rods and cones. Rods are sensitive to dim and achromatic light (night vision), while cones respond to brighter, chromatic light (daylight vision). The retina contains approximately 120 million rods and 7 million cones.

The retina is composed of multiple layers of different cell types (De Valois & De Valois, 1988). Surprisingly, the "inverted" retina is constructed in such a way that photoreceptors are found at the bottom layer. This construction is somewhat counterintuitive since rods and cones are furthest away from incoming light, buried beneath a layer of cells. The retina resembles a three-layer cell sandwich, with connection bundles between each layer. These connectional layers are called plexiform or synaptic layers. The retinogeniculate organization is schematically depicted in Figure 2.4. The outermost layer (w.r.t. incoming light) is the outer nuclear layer which contains the photoreceptor (rod/cone) cells. The first connectional layer is the outer plexiform layer which houses connections between receptor and bipolar nuclei. The next outer layer of cells is the inner nuclear layer containing bipolar (amacrine, bipolar, horizontal) cells. The next plexiform layer is the inner plexiform layer where connections between inner nuclei cells and ganglion cells are formed. The top layer, or the ganglion layer, is composed of ganglion cells.

The fovea's photoreceptors are special types of neurons—the nervous system's basic elements (see Figure 2.5). Retinal rods and cones are specific types of dendrites. In general, individual neurons can connect to as many as 10,000 other neurons. Comprised of such interconnected building blocks, as a whole, the nervous system behaves like a large neural circuit. Certain neurons (e.g., ganglion cells) resemble a "digital gate", sending a signal (firing) when the cell's activation level exceeds a threshold. The myelin sheath is an axonal cover providing insulation which speeds up conduction of impulses. Unmyelinated axons of the ganglion cells converge to the optic disk (an opaque myelin sheath would block light). Axons are myelinated at the optic disk, and connect to the Lateral Geniculate Nuclei (LGN) and the Superior Colliculus (SC).

2.2.1 The Outer Layer

Rods and cones of the outer retinal layer respond to incoming light. A simplified account of the function of these cells is that rods provide monochromatic, scotopic (night) vision, and cones provide trichromatic, photopic (day) vision. Both types of cells are partially sensitive to mesopic (twilight) light levels.

2.2.2 The Inner Nuclear Layer

Outer receptor cells are laterally connected to the horizontal cells. In the fovea, each horizontal cell is connected to about 6 cones, and in the periphery to

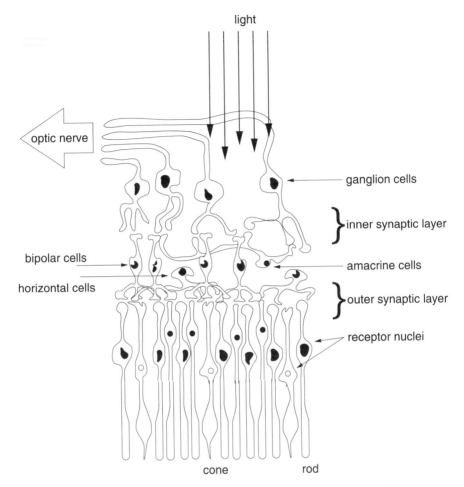

light

optic nerve

ganglion cells

}inner synaptic layer

bipolar cells

amacrine cells

horizontal cells

}outer synaptic layer

receptor nuclei

cone   rod

Fig. 2.4. Schematic diagram of the neural interconnections among receptors and bipolar, ganglion, horizontal, and amacrine cells. Adapted from Dowling and Boycott (1966) with permission © 1966 The Royal Society (London).

about 30-40 cones. Centrally, the cone bipolar cells contact one cone directly, and several cones indirectly through horizontal or receptor-receptor coupling. Peripherally, cone bipolar cells directly contact several cones. The number of receptors increases eccentrically. The rod bipolar cells contact a considerably larger number of receptors than cone bipolars. There are two main types of bipolar cells, ones that depolarize to increments of light (+), and others that depolarize to decrements of light (-). The signal profile (cross-section) of bipolar receptive fields is a "Mexican Hat", or center-surround, with an on-center, or off-center signature.

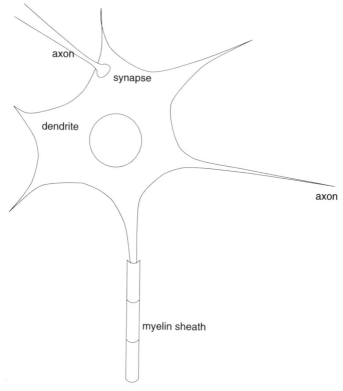

Fig. 2.5. Schematic of the neuron. From BRAIN, MIND, and BEHAVIOR by Floyd E. Bloom and Arlyne Lazerson © 1985, 1988, 2001 by Educational Broadcasting Corporation. Used with the permission of Worth Publishers.

2.2.3 The Ganglion Layer

In a naive view of the Human Visual System (HVS), it is possible to inaccurately think of the retina (and thus the HVS as a whole) acting in a manner similar to that of a camera. While it is true that light enters the eye and is projected through the lens onto the retina, the camera analogy is only accurate up to this point. In the retina, ganglion cells form an "active contrast-enhancing system," not a camera-like plate. Centrally, ganglion cells directly contact one bipolar. Peripherally, ganglion cells directly contact several bipolars. Thus the retinal "camera" is not composed of individual "pixels". Rather, unlike isolated pixels, the retinal photoreceptors (rods and cones in the base layer) form rich interconnections beyond the retinal outer layer. With about 120 million rods and cones and only about 1 million ganglion cells eventually innervating at the LGN, there is considerable convergence of photoreceptor output. That is, the signal of many (on the order of about 100) photoreceptors are combined to produce 1 type of signal. This interconnecting arrangement is described in terms of receptive fields, and this arrangement functions quite differently from a camera.

Ganglion cells are distinguished by their morphological and functional characteristics. Morphologically, there are two types of ganglion cells, the α and β cells. Approximately 10% of retinal ganglion cells are α cells possessing large cell bodies and dendrites, and about 80% are β cells with small bodies and dendrites (Lund, Wu, & Levitt, 1995). The α cells project to the magnocellular (M-) layers of the LGN while the β cells project to the parvocellular (P-) layers. A third channel of input relays through narrow, cell-sparse laminae between the main M- and P-layers of the LGN. Its origin in the retina is not yet known. Functionally, ganglion cells fall into three classes, the X, Y, and W cells (De Valois & De Valois, 1988; Kaplan, 1991). X cells respond to sustained stimulus, location and fine detail, and innervate along both M- and P-projections. Y cells innervate only along the M-projection, and respond to transient stimulus, coarse features, and motion. W cells respond to coarse features, and motion, and project to the Superior Colliculus.

The receptive fields of ganglion cells are similar to those of bipolar cells (center-surround, on-center, off-center). Center-on and center-off receptive fields are depicted in Figure 2.6. Plus signs (+) denote illumination stimulus, minus signs (−) denote lack of stimulus. The vertical bars below each receptive field depict the firing response of the receptive field. This signal characteristic (series of "ticks") is usually obtained by inserting an electrode into the brain. The signal profile of receptive fields resembles the "Mexican hat" operator, often used in image processing.

2.3 The Optic Tract and M/P Visual Channels

Some (but not all) neural signals are transmitted from the retina to the occipital (visual) cortex through the optic tract, crossing in the optic chiasm, making connections to the LGN along the way. The physiology of the optic tract is often described functionally in terms of visual pathways, with reference to specific cells (e.g., ganglion cells). It is interesting to note the decussation (crossing) of the fibers from the nasal half of the retina at the optic chiasm, i.e., nasal retinal signals cross, temporal signals do not.

M and P ganglion cells in the retina connect to M and P channels, respectively. Along the optic pathways, the Superior Colliculus and the Lateral Geniculate Nucleus (LGN) are of particular importance. The SC is involved in program-

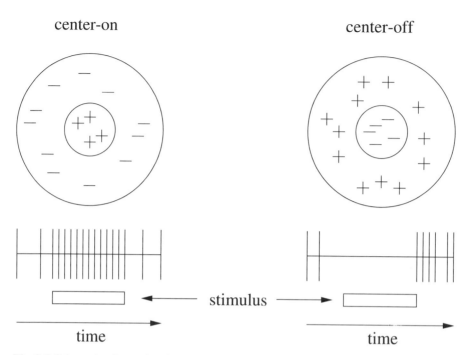

Fig. 2.6. Schematic of receptive-fields.

ming eye movements and also remaps auditory space into visual coordinates. As shown in Figure 2.1, some neural signals along the optic tract project to the SC. The SC is thought to be responsible for directing the eyes to a new region of interest for subsequent detailed visual inspection. Like other regions in the thalamus serving similar functions, the LGN is a cross-over point, or relay station, for α and β ganglion cells. The physiological organization of the LGN, with respect to nervations of these cells, produces a visual field topography of great clinical importance. Here, the Magno-Cellular and the Parvo-Cellular ganglionic projections are clearly visible (under microscope), forming junctions within two distinct layers of the LGN, correspondingly termed the M- and P-layers. Thalamic axons from the M- and P-layers of the LGN terminate in area V1 (the primary visual center) of the striate cortex.

The functional characteristics of ganglionic projections to the LGN and the corresponding Magno- and Parvo-cellular pathways are summarized in Table 2.1. The parvocellular pathway in general responds to signals possessing the following attributes: high contrast (the parvocellular pathway is less sensitive to luminance), chromaticity, low temporal frequency, and high spatial frequency (due to the small receptive fields). Conversely, the magnocellular pathway can be characterized by sensitivity to the following signals: low contrast (the magnocellular pathway is more sensitive to luminance), achromatic-ity, moderate-to-high temporal frequency (e.g., sudden onset stimuli), and low spatial frequency (due to the large receptive fields). Zeki (1993) suggests the existence of four functional pathways defined by the M and P channels: mo-tion, dynamic form, color, and form (size and shape). It is thought that fibers reaching the Superior Colliculus represent retinal receptive fields in rod-rich peripheral zones, while the fibers reaching the LGN represent cone-rich areas of high acuity (Bloom & Lazerson, 1988). It seems likely that, in a general sense, the M ganglion cells correspond to rods, mainly found in the periphery, and the P cells correspond to cones, which are chromatic cells concentrated mainly in the foveal region.

Table 2.1. Functional characteristics of ganglionic projections.

Characteristics	Magnocellular	Parvocellular
ganglion size	large	small
transmission time	fast	slow
receptive fields	large	small
sensitivity to small objects	poor	good
sensitivity to change in light levels	large	small
sensitivity to contrast	low	high
sensitivity to motion	high	low
color discrimination	no	yes

2.4 The Occipital Cortex and Beyond

Thalamic axons from the M- and P-layers of the LGN terminate mainly in the lower and upper halves (β, α divisions, respectively) of layer 4C in middle depth of area V1 (Lund et al, 1995). Cell receptive field size and contrast sensi-tivity signatures are distinctly different in the M- and P- inputs of the LGN, and vary continuously through the depth of layer 4C. Unlike the center-surround receptive fields of retinal ganglion and LGN cells, cortical cells respond to orientation-specific stimulus (Hubel, 1988). Cortical cells are distinguished by two classes: *simple* and *complex*.

In area V1, the size of a simple cell's receptive field depends on its relative retinal position. The smallest fields are in and near the fovea, with sizes of about $1/4 \times 1/4$ degree. This is about the size of the smallest diameters of the smallest receptive field centers of retinal ganglion or LGN cells. In the far periphery, simple cell receptive field sizes are about 1×1 degree. The relationship between small foveal receptive fields and large peripheral receptive fields is maintained about everywhere along the visual pathway.

Simple cells fire only when a line or edge of preferred orientation falls within a particular location of the cell's receptive field. Complex cells fire wherever such a stimulus falls into the cell's receptive field (Lund et al, 1995). The optimum stimulus width for either cell type is, in the fovea, about 2 minutes of arc. The resolving power (acuity) of both cell types is the same.

About 10-20% of complex cells in the upper layers of the striate cortex show marked directional selectivity (Hubel, 1988). Directional selectivity (DS) refers to the cell's response to a particular direction of movement. Cortical directional selectivity (CDS) contributes to motion perception and to the control of eye movements (Grzywacz & Norcia, 1995). CDS cells establish a motion pathway from V1 projecting to MT and V2 (which also projects to MT) and to MST. In contrast, there is no evidence that retinal directional selectivity (RDS) contributes to motion perception. RDS contributes to oculomotor responses (Grzywacz, Sernagor, & Amthor, 1995). In vertebrates, it is involved in optokinetic nystagmus, a type of eye movement discussed in Chapter 4.

2.4.1 Motion-Sensitive Single-Cell Physiology

There are two somewhat counterintuitive implications of the visual system's motion-sensitive single-cell organization for perception. First, due to motion sensitive cells, eye movements are never perfectly still but make constant tiny movements called *microsaccades* (Hubel, 1988). The counterintuitive fact regarding eye movements is that if an image were artificially stabilized on the retina, vision would fade away within about a second and the scene would become blank. Second, due to the response characteristics of single (cortical) cells, the camera-like "retinal buffer" representation of natural images is much more abstract than what intuition suggests. An object in the visual field stimulates only a tiny fraction of the cells on whose receptive field it falls (Hubel, 1988). Perception of the object depends mostly on the response of (orientation-specific) cells to the object's borders. For example, the homogeneously shaded interior of an arbitrary form (e.g., a kidney bean) does not stimulate cells of the visual system. Awareness of the interior shade or hue depends on only cells sensitive to the borders of the object. In Hubel's (1988) words, "...our perception of the interior as black, white, gray, or green has nothing to do with cells whose fields are in the interior—hard as that may be to swallow...What happens at the borders is the only information you need to know: the interior is boring."

2.5 Summary and Further Reading

This chapter presented a simplified view of the brain with emphasis on regions and structures of the brain responsible for attentional and visual processing, including those regions implicated in eye movement generation. Starting with the structure of the eye, the most salient observation is the structure of the retina which clearly shows the limited scope of the high resolution fovea. The division between foveo-peripheral vision is maintained along the visual pathways and can be clearly seen under microscope in the LGN. Of particular relevance to attention and eye movements is the physiological and functional duality of the Magno- and Parvo-cellular pathways and of their apparent mapping to their attentional "what" and "where" classification. While this characterization of the M- and P-pathways is admittedly overly simplistic, it provides an intuitive functional distinction between foveal and peripheral vision.

An interesting visual example of foveo-peripheral processing is shown in Figure 2.7. To notice the curious difference between foveal and peripheral processing, foveate one corner of the image in Figure 2.7 and, without moving your eyes, shift your attention to the opposing corner of the image. Interestingly, you should perceive white dots at the line crossings in the foveal region, but black dots should appear at the line crossings in the periphery.

Examining regions in the brain along the visual pathways, one can obtain insight into how the brain processes visual information. The notion that attention may be driven by certain visual features (e.g., edges) is supported to an extent by the identification of neural regions which respond to these features. How certain features are perceived, particularly within and beyond the fovea, is the topic covered in the next chapter.

For an excellent review of physiological optics and visual perception in general, see Hendee and Wells (1997). For an introduction to neuroscience, see Hubel's (1988) very readable text. For a more recent description of the brain with an emphasis on color vision, see Zeki (1993). Apart from these texts on vision, several "handbooks" have also been assembled describing current knowledge of the brain. Arbib's (1995) handbook is one such example. It is an excellent source summarizing current knowledge of the brain, although it is

Fig. 2.7. Foveo-peripheral illusion: scintillation effect produced by a variation of the standard Hermann grid illusion (attributed to L. Hermann (1870)), first discovered by Elke Lingelbach (at home). Adapted from Ninio and Stevens, Variations on the Hermann Grid: An Extinction Illusion, *Perception*, 29, 1209-1217, 2000 © 2000 Pion, London.

somewhat difficult to read and to navigate through. [1] Another such well organized but rather large text is Gazzaniga (2000).

---

[1] A new edition of Arbib's book has recently been announced.

3. Visual Psychophysics

Given the underlying physiological substrate of the Human Visual System, measurable performance parameters often (but not always!) fall within ranges predicted by the limitations of the neurological substrate. Visual performance parameters, such as visual acuity, are often measured following established experimental paradigms, generally derived in the field of psychophysics (e.g., Receiver Operating Characteristics, or ROC paradigm, is one of the more popular experimental methods).

Unexpected observed visual performance is often a consequence of complex visual processes (e.g., visual illusions), or combinations of several factors. For example, the well-known Contrast Sensitivity Function, or CSF, describing the Human Visual System's response to stimuli of varying contrast and resolution, depends not only on the organization of the retinal mosaic, but also on the response characteristics of complex cellular combinations, e.g., receptive fields.

In this book, the primary concern is visual attention, and so the book primarily considers the distinction between foveo-peripheral vision. This subject, while complex, is discussed here in a fairly simplified manner, with the aim of elucidating only the most dramatic differences between what is perceived foveally and peripherally. In particular, visual (spatial) acuity is arguably the most studied distinction and is possibly the simplest parameter to alter in eye-based interaction systems (at least at this time). It is therefore the topic covered in greatest detail, in comparison to the other distinctions covered here briefly: temporal and chromatic foveo-peripheral differences.

3.1 Spatial Vision

Dimensions of retinal features are usually described in terms of projected scene dimensions in units of degrees visual angle, defined as

$$A = 2 \arctan \frac{S}{2D},$$

where S is the size of the scene object and D is the distance to the object (see Figure 3.1). Common visual angles are given in Table 3.1.

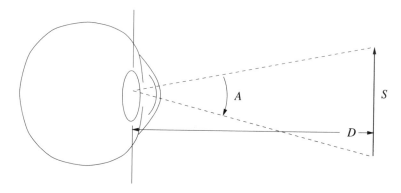

Fig. 3.1. Visual angle. Adapted from THE PSYCHOLOGY OF VISUAL PERCEPTION, *1st edition*, by Haber and Hershenson (1973) © 1973. Reprinted with permission of Brooks/Cole, an imprint of the Wadsworth Group, a division of Thomson Learning. Fax 800 730-2215.

Table 3.1. Common visual angles.

Object	Distance	Angle subtended
thumbnail	arm's length	1.5–2°
sun or moon	–	.5° or 30′ or arc
US quarter coin	arm's length	2°
US quarter coin	85m	1′ (1 minute of arc)
US quarter coin	5km	1″ (1 second of arc)

The innermost region is the fovea centralis (or foveola) which measures 400μm in diameter and contains 25,000 cones. The fovea proper measures 1500μm in diameter and holds 100,000 cones. The macula (or central retina) is 5000μm in diameter, and contains 650,000 cones. One degree visual angle corresponds to approximately 300μm distance on the human retina (De Valois & De Valois, 1988). The foveola, measuring 400μm subtends 1.3° visual angle, while the fovea and macula subtend 5° and 16.7°, respectively (see Figure 3.2). Figure 3.3 shows the retinal distribution of rod and cone receptors. The fovea contains 147,000 cones/mm^2 and a slightly smaller number of rods. At about 10° the number of cones drops sharply to less than 20,000 cones/mm^2 while at 30° the number of rods in the periphery drops to about 100,000 rods/mm^2 (Haber & Hershenson, 1973).

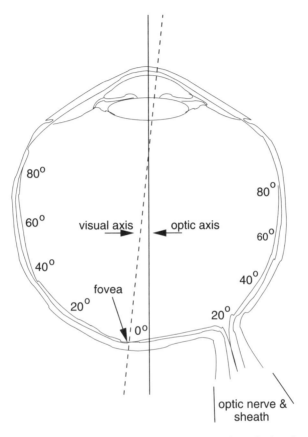

Fig. 3.2. Density distributions of rod and cone receptors across the retinal surface: visual angle. Adapted from Pirenne (1967) (as cited in Haber and Hershenson (1973)).

The entire visual field roughly corresponds to a 23,400 square degree area defined by an ellipsoid with the horizontal major axis subtending 180° visual angle, and the minor vertical axis subtending 130°. The diameter of the highest acuity circular region subtends 2°, the parafovea (zone of high acuity) extends to about 4° or 5°, and acuity drops off sharply beyond. At 5°, acuity is only 50% (Irwin, 1992). The so-called "useful" visual field extends to about 30°. The rest of the visual field has very poor resolvable power and is mostly used for perception of ambient motion. With increasing eccentricity the cones increase in size, while the rods do not (De Valois & De Valois, 1988). Cones, not rods, make the largest contribution to the information going to deeper brain

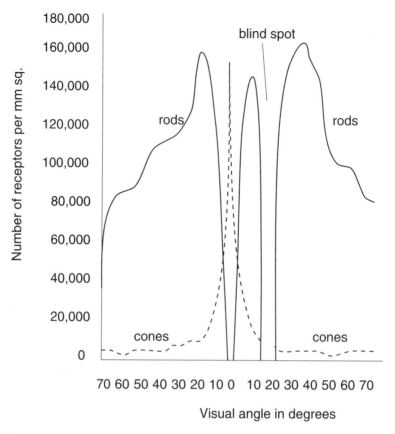

Fig. 3.3. Density distributions of rod and cone receptors across the retinal surface: rod/cone density. Adapted from Pirenne (1967) (as cited in Haber and Hershenson (1973)).

centers, and provide most of the fine-grained spatial resolvability of the visual system.

The Modulation Transfer Function (MTF) theoretically describes the spatial resolvability of retinal photoreceptors by considering the cells as a finite array of sampling units. The 400μm-diameter rod-free foveola contains 25,000 cones. Using the area of a circle, $25000 = \pi r^2$, approximately $2\sqrt{25000/\pi} = 178.41$ cones occupy a 400μm linear cross-section of the foveola with an estimated average linear inter-cone spacing of 2.24μm. Cones in this region measure about 1μm in diameter. Since one degree visual angle corresponds to approximately 300μm distance on the human retina, roughly 133 cones are packed per degree visual angle in the foveola. By the sampling theorem, this suggests a resolvable spatial Nyquist frequency of 66 c/deg. Subjective resolution has in fact been measured at about 60 c/deg (De Valois & De Valois, 1988).

In the fovea, a similar estimate based on the foveal diameter of 1500μm and a 100,000 cone population, gives an approximate linear cone distribution of $2\sqrt{100000/\pi} = 356.82$ cones per 1500μm. The average linear inter-cone spacing is then 71 cones/deg suggesting a maximum resolvable frequency of 35 cycles/deg, roughly half the resolvability within the foveola. This is somewhat of an underestimate since cone diameters increase two-fold by the edge of the fovea suggesting a slightly milder acuity degradation. These one-dimensional approximations are not fully generalizable to the two-dimensional photoreceptor array although they provide insight into the theoretic resolution limits of the eye. Effective relative visual acuity measures are usually obtained through psychophysical experimentation.

At photopic light levels (day, or cone vision), foveal acuity is fairly constant within the central $2°$, and drops approximately linearly from there to the $5°$ foveal border. Beyond the $5°$, acuity drops sharply (approximately exponentially). At scotopic light levels (night, or rod-vision), acuity is poor at all eccentricities. Figure 3.4 shows the variation of visual acuity at various eccentricities and light intensity levels. Intensity is shown varying from 9.0 to 4.6 log micromicrolamberts, denoted by log mmL (9.0 log micromicrolamberts = 10^9 micromicrolamberts = 1 mL, see Davson (1980, p.311)). The correspondence between foveal receptor spacing and optical limits generally holds in foveal regions of the retina, but not necessarily in the periphery. In contrast to the approximate 60 c/deg resolvability of foveal cones, the highest spatial frequencies resolvable by rods are on the order of 5 c/deg, suggesting poor resolvability in the relatively cone-free periphery. Although visual acuity correlates fairly well with cone distribution density, it is important to note that synaptic organization and later neural elements (e.g., ganglion cells concentrated in the central retina) are also contributing factors in determining visual acuity.

3.2 Temporal Vision

Human visual response to motion is characterized by two distinct facts: the *persistence of vision* and the *phi phenomenon* (Gregory, 1990). The former essentially describes the temporal sampling rate of the HVS, while the latter describes a threshold above which the HVS detects *apparent movement*. Both facts are exploited in television, cinema, and graphics to elicit perception of motion from successively displayed still images.

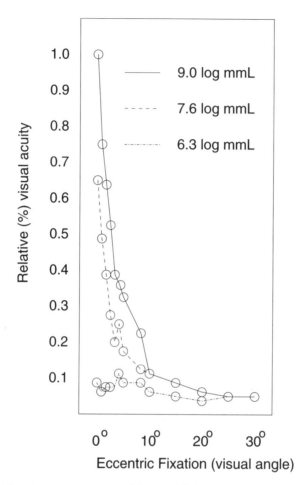

Fig. 3.4. Visual acuity at various eccentricities and light levels. Adapted from Davson (1980) with permission © 1980 Academic Press.

Persistence of vision describes the inability of the retina to sample rapidly changing intensities. A stimulus flashing at about 50-60Hz (cycles per second) will appear steady (depending on contrast and luminance conditions and observers). This is known as the Critical Fusion Frequency (CFF).[1] A stylized representation of the CFF, based on measurements of response to temporal stimuli of varying contrast, i.e., a temporal contrast sensitivity function, is shown in Figure 3.5. Incidentally, the curve of the CFF resembles the shape of the curve of the Contrast Sensitivity Function (CSF) which describes retinal spatial frequency response. The CFF explains why flicker is not seen when viewing a sequence of (still) images at a high enough rate. The CFF illusion

---

[1] Also sometimes referred to as the Critical Flicker Frequency.

is maintained in cinema since frames are shown at 24 frames per second (fps, equivalent to Hz), but a three-bladed shutter raises the flicker rate to 72Hz (three for each picture). Television also achieves the CFF by displaying the signal at 60 fields per second. Television's analog to cinema's three-bladed shutter is the interlacing scheme: the typical television frame rate is about 30 frames per second (depending on the standard use, e.g., NTSC in North America, PAL in other regions), but only the even or odd scanlines (fields) are shown per cycle. Although the CFF explains why flicker is effectively eliminated in motion picture (and computer) displays, it does not fully explain why motion is perceived.

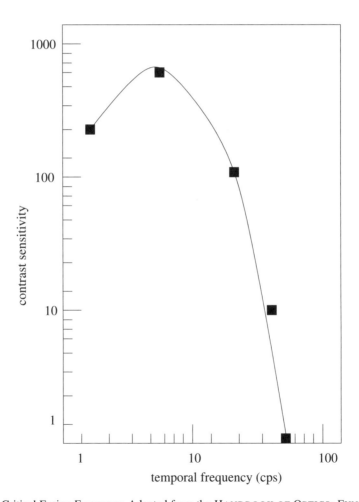

Fig. 3.5. Critical Fusion Frequency. Adapted from the HANDBOOK OF OPTICS: FUNDAMEN-TALS, TECHNIQUES, & DESIGN, *2nd edition*, by Bass (1995) © 1995 McGraw-Hill. Reproduced with permission of The McGraw-Hill Companies.

The second fact that explains why movies, television, and graphics work is the phi phenomenon, or *stroboscopic motion*, or *apparent motion*. This fact explains the illusion of old-fashioned moving neon signs whose stationary lights are turned on in quick succession. This illusion can also be demonstrated with just two lights, provided the delay between successive light flashes is no less than about 62Hz (Brinkmann, 1999). Inverting this value gives a rate of about 16fps which is considered a bare minimum to facilitate the illusion of apparent motion.

3.2.1 Perception of Motion in the Visual Periphery

In the context of visual attention and foveo-peripheral vision, the temporal response of the HVS is not homogeneous across the visual field. In terms of motion responsiveness, Koenderink, van Doorn, and van de Grind (1985) provide support that the foveal region is more receptive to slower motion than the periphery, although motion is perceived uniformly across the visual field. Sensitivity to target motion decreases monotonically with retinal eccentricity for slow and very slow motion (1 cycle/deg) (Boff & Lincoln, 1988). That is, the velocity of a moving target appears slower in the periphery than in the fovea. Conversely, a higher rate of motion (e.g., frequency of rotation of grated disk) is needed in the periphery to match the apparent stimulus velocity in the fovea. At higher velocities, the effect is reversed.

Despite the decreased sensitivity in the periphery, movement is more salient there than in the central field of view (fovea). That is, the periphery is more sensitive to moving targets than to stationary ones. It is easier to peripherally detect a moving target than it is a stationary one. In essence, motion detection is the periphery's major task; it is a kind of early warning system for moving targets entering the visual field.

3.2.2 Sensitivity to Direction of Motion in the Visual Periphery

The periphery is approximately twice as sensitive to horizontal-axis movement as to vertical-axis movement (Boff & Lincoln, 1988). Directional motion sensitivity is show in Figure 3.6.

3.3 Color Vision

Foveal color vision if facilitated by the three types of retinal cone photoreceptors. The three main spectral sensitivity curves for retinal cone photoreceptors

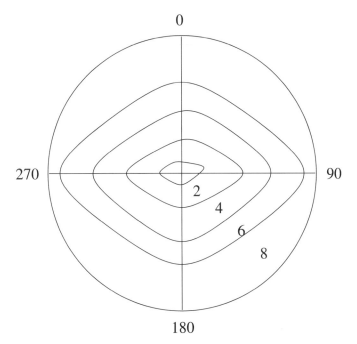

Fig. 3.6. Absolute threshold isograms for detecting peripheral rotary movement. Numbers are rates of pointer movement in revolutions per minute. Adapted from McColgin (1960) with permission © 1960 Optical Society of America.

peak at approximately 450 nm, 520 nm, and 555 nm wavelengths, for each of the blue, green, and red photoreceptors, respectively. A great deal is known about color vision in the fovea, however, relatively little is known about peripheral color vision. Of the 7 million cones, most are packed tightly into the central 30° region of the fovea with scarcely any cones found beyond. This cone distribution suggests that peripheral color vision is quite poor in comparison to the color sensitivity of the central retinal region. Visual fields for monocular color vision are shown in Figure 3.7. Fields are shown for the right eye; fields for the left eye would be mirror images of those for the right eye. Blue and yellow fields are larger than the red and green fields; no chromatic visual fields have definite border, instead, sensitivity drops off gradually and irregularly over a range of 15-30° visual angle (Boff & Lincoln, 1988).

Quantification of perceptual performance is not easily found in the literature. Compared to investigation of foveal color vision, only a few experiments have been performed to measure peripheral color sensitivity. Two studies, of par-

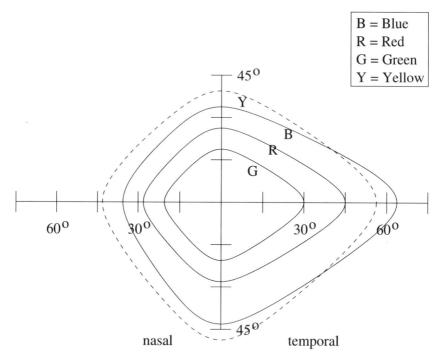

Fig. 3.7. Visual fields for monocular color vision (right eye). Adapted from Boff and Lincoln (1988) with permission ©1988 Wright-Patterson AFB.

ticular relevance to peripheral location of color CRTs in an aircraft cockpit environment, investigated the chromatic discrimination of peripheral targets.

In the first study, Doyal (1991) concludes that peripheral color discrimination can approximate foveal discrimination when relatively small field sizes are presented (e.g., 2° at 10° eccentricity, and less than 4° at 25°). While this sounds encouraging, color discrimination was tested at limited peripheral eccentricities (within the central 30°).

In the second, Ancman (1991) tested color discrimination at much greater eccentricities, up to about 80° visual angle. She found that subjects wrongly identified the color of a peripherally located 1.3° circle displayed on a CRT 5% of the time if it was blue, 63% of the time if red, and 62% of the time if green. Furthermore, blue could not be seen further than 83.1°off the fovea (along the x-axis); red had to be closer than 76.3° and green nearer than 74.3° before subjects could identify the color.

There is much yet to be learned about peripheral color vision. Being able to verify a subject's direction of gaze during peripheral testing would be of significant benefit to these experiments. This type of psychophysical testing is but one of several research areas where eye tracking studies could play an important supporting role.

3.4 Implications for Attentional Design of Visual Displays

Both the structure and functionality of human visual system components place constraints on the design parameters of a visual communication system. In particular, the design of a gaze-contingent system must distinguish the characteristics of foveal and peripheral vision (see Section 14.2). A *visuotopic* representation model for imagery based on these observations is proposed:

1. **Spatial Resolution** should remain high within the foveal region and smoothly degrade within the periphery, matching human visual acuity. High spatial frequency features in the periphery must be made visible "just in time" to anticipate gaze-contingent fixation changes.
2. **Temporal Resolution** must be available in the periphery. Sudden onset events are potential attentional attractors. At low speeds, motion of peripheral targets should be increased to match apparent motion in the central field of view.
3. **Luminance** should be coded for high visibility in the peripheral areas since the periphery is sensitive to dim objects.
4. **Chrominance** should be coded for high exposure almost exclusively in the foveal region, with chromaticity decreasing sharply into the periphery. This requirement is a direct consequence of the high density of cones and parvocellular ganglion cells in the fovea.
5. **Contrast** sensitivity should be high in the periphery, corresponding to the sensitivity of the magnocellular ganglion cells found mainly outside the fovea.

Special consideration should be given to sudden onset, luminous, high frequency objects (i.e., suddenly appearing bright edges).

A gaze-contingent visual system faces an implementational difficulty not yet addressed: matching the dynamics of human eye movement. Any system designed to incorporate an eye-slaved high resolution of interest, for example, must deal with the inherent delay imposed by the processing required to track and process real-time eye tracking data. To consider the temporal constraints that need to be met by such systems, the dynamics of human eye movements must be evaluated. This topic is considered in the following chapter.

3.5 Summary and Further Reading

Psychophysical information may be the most usable form of literature for the design of graphical displays, attentional in nature or otherwise. Introductory texts may include function plots of some aspect of vision (e.g., acuity) which may readily be used to guide the design of visual displays. However, one often needs to evaluate the experimental design used in psychophysical experiments to determine the generalizability of reported results. Furthermore, similar caution should be employed as in reading neurological literature: psychophysical results may often deal with a certain specific aspect of vision, which may or may not be readily applicable to display design. For example, visual acuity may suggest the use of relatively sized fonts on a web page (larger font in the periphery), but acuity alone may not be sufficient to determine the required resolution in something like an attentional image or video display program. For the latter, one may need to piece together information concerning the visual contrast sensitivity function, temporal sensitivity, etc. Furthermore, psychophysical studies may involve relatively simple stimuli (sine wave gratings), the results of which may or may not generalize to more complex stimuli such as imagery.

For a good introductory book on visual perception, see Hendee and Wells (1997). This text includes a good introductory chapter on the neurological basis of vision. Another good introductory book which also includes an interesting perspective on the perception of art is Solso (1999). For a somewhat terse but fairly complete psychophysical reference, see the USAF Engineering Data Compendium (Boff & Lincoln, 1988). This is an excellent, "quick" guide to visual performance.

4. Taxonomy and Models of Eye Movements

Almost all normal primate eye movements used to reposition the fovea result as combinations of five basic types: saccadic, smooth pursuit, vergence, vestibular, and physiological nystagmus (miniature movements associated with fixations) (Robinson, 1968). Vergence movements are used to focus the pair of eyes over a distant target (depth perception). Other movements such as adaptation and accommodation refer to non-positional aspects of eye movements (i.e., pupil dilation, lens focusing). With respect to visual display design, positional eye movements are of primary importance.

4.1 The Extra-Ocular Muscles and The Oculomotor Plant

In general, the eyes move within six degrees of freedom: three translations within the socket, and three rotations. There are six muscles responsible for movement of the eyeball: the *medial* and *lateral recti* (sideways movements), the *superior* and *inferior recti* (up/down movements), and the *superior* and *inferior obliques* (twist) (Davson, 1980). These are depicted in Figure 4.1. The neural system involved in generating eye movements is known as the oculomotor plant. The general plant structure and connections are shown in Figure 4.2 and described in Robinson (1968). Eye movement control signals emanate from several functionally distinct regions. Areas 17, 18, 19, and 22 are areas in the occipital cortex thought to be responsible for high-level visual functions such as recognition. The superior colliculus bears afferents emanating directly from the retina, particularly from peripheral regions conveyed through the magno-cellular pathway. The semicircular canals react to head movements in three-dimensional space. All three areas, i.e., the occipital cortex, the superior colliculus, and the semicircular canals convey efferents to the eye muscles through the mesencephalic and pontine reticular formations. Classification of observed eye movement signals relies in part on the known functional characteristics of these cortical regions.

Left (view from above): 1, superior rectus; 2, levator palbebrae superioris; 3, lateral rectus; 4, medial rectus; 5, superior oblique; 6, reflected tendon of the superior oblique; 7, annulus of Zinn. *Right (lateral view):* 8, inferior rectus; 9, inferior oblique.

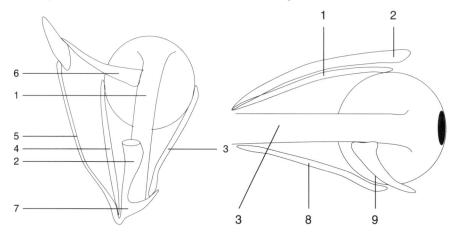

Fig. 4.1. Extrinsic muscles of the eye. Adapted from Davson (1980) with permission © 1980 Academic Press.

Two pertinent observations regarding eye movements can be drawn from the oculomotor plant's organization:

1. The eye movement system is, to a large extent, a feedback circuit.
2. Signals controlling eye movement emanate from cortical regions which can be functionally categorized as voluntary (occipital cortex), involuntary (superior colliculus), and reflexive (semicircular canals).

The feedback-like circuitry is utilized mainly in the types of eye movements requiring stabilization of the eye. Orbital equilibrium is necessitated for the steady retinal projection of an object, concomitant with the object's motion and movements of the head. Stability is maintained by a neuronal control system.

4.2 Saccades

Saccades are rapid eye movements used in repositioning the fovea to a new location in the visual environment. The term comes from an old French word meaning "flick of a sail" (Gregory, 1990). Saccadic movements are both voluntary and reflexive. The movements can be voluntarily executed or they can be invoked as a corrective optokinetic or vestibular measure (see below). Saccades range in duration from 10ms to 100ms, which is a sufficiently short duration to render the executor effectively blind during the transition (Shebilske & Fisher, 1983). There is some debate over the underlying neuronal

CBT, corticobular tract; CER, cerebellum; ICTT, internal corticotectal tract; LG, lateral geniculate body; MLF, medial longitudinal fasciculus; MRF, mesencephalic and pontine reticular
formations; PT, pretectal nuclei; SA, stretch afferents from extraocular muscles; SC, superior
colliculi; SCC, semicircular canals; T, tegmental nuclei; VN, vestibular nuclei; II, optic nerve;
III, IV, and VI, the oculomotor, trochlear, and abducens nuclei and nerves; 17, 18, 19, 22, primary and association visual areas, occipital and parietal (Brodmann); 8, the frontal eye fields.

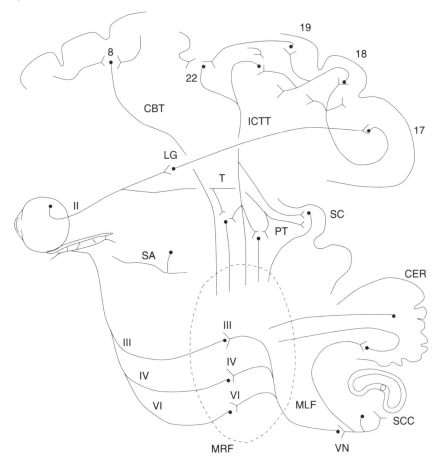

Fig. 4.2. Schematic of the major known elements of the oculomotor system. Adapted from
Robinson (1968) with permission © 1968 IEEE.

system driving saccades. On the one hand, saccades have been deemed ballistic
and stereotyped. The term stereotyped refers to the observation that particular
movement patterns can be evoked repeatedly. The term ballistic refers to the
presumption that saccade destinations are pre-programmed. That is, once the
saccadic movement to the next desired fixation location has been calculated
(programming latencies of about 200ms have been reported), saccades cannot
be altered. One reason behind this presumption is that, during saccade execu-

tion, there is insufficient time for visual feedback to guide the eye to its final position (Carpenter, 1977). One the other hand, a saccadic feedback system is plausible if it is assumed that instead of visual feedback, an internal copy of head, eye, and target position is used to guide the eyes during a saccade (Laurutis & Robinson, 1986; Fuchs, Kaneko, & Scudder, 1985). Due to their fast velocities, saccades may only appear to be ballistic (Zee, Optican, Cook, Robinson, & Engel, 1976).

Various models for saccadic programming have been proposed (Findlay, 1992). These models, with the exception of ones including "center-of-gravity" coding (see for example He and Kowler (1989)), may inadequately predict unchangeable saccade paths. Instead, saccadic feedback systems based on an internal representation of target position may be more plausible since they tend to correctly predict the so-called double-step experimental paradigm. The double-step paradigm is an experiment where target position is changed during a saccade in mid-flight. Fuchs et al (1985) proposed a refinement of Robinson's feedback model which is based on a signal provided by the superior colliculus and a local feedback loop. The local loop generates feedback in the form of motor error produced by subtracting eye position from a mental target-in-space position. Sparks and Mays (1990) cite compelling evidence that intermediate and deep layers of the SC contain neurons that are critical components of the neural circuitry initiating and controlling saccadic movements. These layers of the SC receive inputs from cortical regions involved in the analysis of sensory (visual, auditory, and somatosensory) signals used to guide saccades. The authors also rely on implications of Listing's and Donder's Laws which specify an essentially null torsion component in eye movements, requiring virtually only two degrees of freedom for saccadic eye motions (Davson, 1980; Sparks & Mays, 1990). According to these laws, motions can be resolved into rotations about the horizontal x- and vertical y-axes.

Models of saccadic generation attempt to provide an explanation of the underlying mechanism responsible for generating the signals sent to the motor neurons. Although there is some debate as to the source of the saccadic program, the observed signal resembles a pulse/step function (Sparks & Mays, 1990). The pulse/step function refers to a dual velocity and position command to the extraocular muscles (Leigh & Zee, 1991). A possible simple representation of a saccadic step signal is a differentiation filter. Carpenter (1977) suggests such a possible filter arrangement for generating saccades coupled with an integrator. The integrating filter is in place to model the necessary conversion of velocity-coded information to position-coded signals (Leigh & Zee, 1991).

A perfect neural integrator converts a pulse signal to a step function. An imperfect integrator (called leaky) will generate a signal resembling a decaying exponential function. The principle of this type of neural integration applies to all types of conjugate eye movements. Neural circuits connecting structures in the brain stem and the cerebellum exist to perform integration of coupled eye movements including saccades, smooth pursuits, and vestibular and optokinetic nystagmus (see below) (Leigh & Zee, 1991).

A differentiation filter can be modeled by a linear filter as shown in Figure 4.3. In the time domain, the linear filter is modeled by the following equation

$$x_t = g_0 s_t + g_1 s_{t-1} + \cdots$$
$$= \sum_{k=0}^{\infty} g_k s_{t-k},$$

where s_t is the input (pulse), x_t is the output (step), and g_k are the filter coefficients. To ensure differentiation, the filter coefficients typically must satisfy properties which approximate mathematical differentiation. An example of such a filter is the Haar filter with coefficients $\{1, -1\}$. Under the z-transform the transfer function $X(z)/S(z)$ of this linear filter is

$$x_t = g_0 s_t + g_1 s_{t-1}$$
$$x_t = (1)s_t + (-1)s_{t-1}$$
$$x_t = (1)s_t + (-1)z s_t$$
$$x_t = (1-z)s_t$$
$$X(z) = (1-z)S(z)$$
$$\frac{X(z)}{S(z)} = 1 - z.$$

The Haar filter is a length-2 filter which approximates the first derivate between successive pairs of inputs.

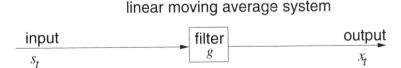

linear moving average system

| input | filter | output |

s_t · · · g · · · x_t

Fig. 4.3. Diagram of a simple linear filter modeling saccadic movements.

4.3 Smooth Pursuits

Pursuit movements are involved when visually tracking a moving target. Depending on the range of target motion, the eyes are capable of matching the velocity of the moving target. Pursuit movements provide an example of a control system with built-in negative feedback (Carpenter, 1977). A simple closed-loop feedback loop used to model pursuit movements is shown in Figure 4.4, where s_t is the target position, x_t is the (desired) eye position, and h is the (linear, time-invariant) filter, or gain of the system (Carpenter, 1977; Leigh & Zee, 1991). Tracing the loop from the feedback start point gives the following equation in the time domain

$$h(s_t - x_t) = x_{t+1}.$$

Under the z-transform the transfer function $X(z)/S(z)$ of this linear system is

$$H(z)(S(z) - X(z)) = X(z)$$
$$H(z)S(z) = X(z)(1 + H(z))$$
$$\frac{H(z)}{1 + H(z)} = \frac{X(z)}{S(z)}.$$

In the closed-loop feedback model, signals from visual receptors constitute the error signal indicating needed compensation to match the target's retinal image motion.

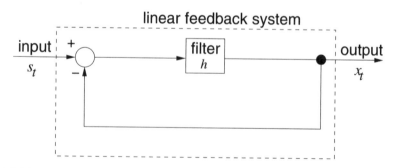

Fig. 4.4. Diagram of a simple linear feedback model of smooth pursuit movements.

4.4 Fixations

Fixations are eye movements which stabilize the retina over a stationary object of interest. It seems intuitive that fixations should be generated by the same neuronal circuit controlling smooth pursuits with fixations being a special case of a target moving at zero velocity. This is probably incorrect (Leigh & Zee, 1991, pp.139-140). Fixations, instead, are characterized by the miniature eye movements: tremor, drift, and microsaccades. This is a somewhat counterintuitive consequence of the visual system's motion-sensitive single-cell organization. Recall that microsaccades are made due to the motion sensitivity of the visual system's single-cell physiology. Microsaccades are eye movement signals that are more or less spatially random varying over 1 to 2 minutes of arc in amplitude. The counterintuitive fact regarding fixations is that if an image is artificially stabilized on the retina, vision fades away within about a second and the scene becomes blank.

Miniature eye movements that effectively characterize fixations may be considered noise present in the control system (possibly distinct from the smooth pursuit circuit) attempting to hold gaze steady. This noise appears as a random fluctuation about the area of fixation, typically no larger than 5° visual angle (Carpenter, 1977, p.105). Although the classification of miniature movements as noise may be an oversimplification of the underlying natural process, it allows the signal to be modeled by a feedback system similar to the one shown in Figure 4.4. The additive noise in Figure 4.4 is represented by $e_t = s_t - x_t$, where the (desired) eye position x_t is subtracted from the steady fixation position s_t at the summing junction. In this model, the error signal stimulates the fixation system in a manner similar to the smooth pursuit system, except that here e_t is an error-position signal instead of an error-velocity signal (see Leigh and Zee (1991, p.150)). The feedback system modeling fixations, using the noisy "data reduction" method, is in fact simpler than the pursuit model since it implicitly assumes a stationary stochastic process (Carpenter, 1977, p.107). Stationarity in the statistical sense refers to a process with constant mean. Other relevant statistical measures of fixations include their duration range of 150ms to 600ms, and the observation that 90% of viewing time is devoted to fixations (Irwin, 1992).

4.5 Nystagmus

Nystagmus eye movements are conjugate eye movements characterized by a sawtooth-like time course (time series signal) pattern. Optokinetic nystagmus is a smooth pursuit movement interspersed with saccades invoked to compensate for the retinal movement of the target. The smooth pursuit component of optokinetic nystagmus appears in the slow phase of the signal (Robinson, 1968). Vestibular nystagmus is a similar type of eye movement compensating for the movement of the head. The time course of vestibular nystagmus is virtually indistinguishable from its optokinetic counterpart (Carpenter, 1977).

4.6 Implications for Eye Movement Analysis

From the above discussion, two significant observations relevant to eye movement analysis can be made. First, based on the functionality of eye movements, only three types of movements need be modeled to gain insight into the overt localization of visual attention. These types of eye movements are fixations, smooth pursuits, and saccades. Second, based on signal characteristics and plausible underlying neural circuitry, all three types of eye movements may be approximated by a linear, time-invariant (LTI) system (i.e., a linear filter– for examples of linear filters applicable to saccade detection, see Chapter 9).

The primary requirement of eye movement analysis, in the context of gaze-contingent system design, is the identification of fixations, saccades, and smooth pursuits. It is assumed that these movements provide evidence of voluntary, overt visual attention. This assumption does not preclude the plausible involuntary utility of these movements, or conversely, the covert non-use of these eye movements (e.g., as in the case of parafoveal attention). Fixations naturally correspond to the desire to maintain one's gaze on an object of interest. Similarly, pursuits are used in the same manner for objects in smooth motion. Saccades are considered manifestations of the desire to voluntarily change the focus of attention.

4.7 Summary and Further Reading

This chapter presented a taxonomy of eye movements and included linear models of eye movement signals suitable for eye movement analysis (see also Chapter 9.

With the exception of Carpenter's widely referenced text (Carpenter, 1977), there appears to be no single suitable introductory text discussing eye movements exclusively. Instead, there are various texts on perception, cognition, and neuroscience which often include a chapter or section on the topic. There are also various collections of technical papers on eye movements, usually assembled from proceedings of focused symposia or conferences. A series of such books was produced by John Senders et al. in the 1970s and 1980s (see for example Monty and Senders (1976), Fisher, Monty, and Senders (1981)). This conference series has recently been revived in the form of the (currently biennial) Eye Tracking Research & Applications (ETRA) conference).

A large amount of work has been performed on studying eye movements in the context of reading. For a good introduction to this literature, see Rayner (1992).

Part II

Eye Tracking Systems

5. Eye Tracking Techniques

The measurement device most often used for measuring eye movements is commonly known as an eye tracker. In general, there are two types of eye movement monitoring techniques: those that measure the position of the eye relative to the head, and those that measure the orientation of the eye in space, or the "point of regard" (Young & Sheena, 1975). The latter measurement is typically used when the concern is the identification of elements in a visual scene, e.g., in (graphical) interactive applications. Possibly the most widely applied apparatus for measurement of the point of regard is the video-based corneal reflection eye tracker. In this chapter, most of the popular eye movement measurement techniques are briefly discussed first before covering video-based trackers in greater detail.

There are four broad categories of eye movement measurement methodologies involving the use or measurement of: electro-oculography (EOG), scleral contact lens/search coil, photo-oculography (POG) or video-oculography (VOG), and video-based combined pupil and corneal reflection.

Electro-oculography, or EOG, relies on (d.c. signal) recordings of the electric potential differences of the skin surrounding the ocular cavity. During the mid-1970s, this technique was the most widely applied eye movement method (Young & Sheena, 1975). Today, possibly the most widely applied eye movement technique, primarily used for point of regard measurements, is the method based on corneal reflection.

The first method for objective eye measurements using corneal reflection was reported in 1901 (Robinson, 1968). To improve accuracy, techniques using a contact lens were developed in the 1950s. Devices attached to the contact lens ranged from small mirrors to coils of wire. Measurement devices relying on physical contact with the eyeball generally provide very sensitive measurements. The obvious drawback of these devices is their invasive requirement of wearing the contact lens. So-called non-invasive (sometimes called remote) eye trackers typically rely on the measurement of visible features of the eye, e.g., the pupil, iris-sclera boundary, or a corneal reflection of a closely positioned, directed light source. These techniques often involve either manual or automatic (computer-based) analysis of video recordings of the movements of the eyes, either off-line or in real-time. The availability of fast image processing hardware has facilitated the development of real-time video-based point of regard turn-key systems.

5.1 Electro-Oculography (EOG)

Electro-oculography, or EOG, the most widely applied eye movement recording method some forty years ago (and still used today), relies on measurement of the skin's electric potential differences, of electrodes placed around the eye. A picture of a subject wearing the EOG apparatus is shown in Figure 5.1. The recorded potentials are in the range $15\text{-}200\mu V$, with nominal sensitivities of order of $20\mu V/\text{deg}$ of eye movement. This technique measures eye movements relative to head position, and so is not generally suitable for point of regard measurements unless head position is also measured (e.g., using a head tracker).

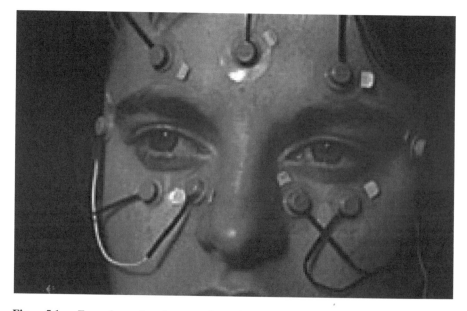

Fig. 5.1. Example of electro-oculography (EOG) eye movement measurement. Courtesy of MetroVision, 4 rue des Platanes, 59840 Pérenchies, France <http://www.metrovision.fr>. Reproduced with permission.

5.2 Scleral Contact Lens/Search Coil

One of the most precise eye movement measurement methods involves attaching a mechanical or optical reference object mounted on a contact lens which is then worn directly on the eye. Such early recordings (ca. 1898 (Young & Sheena, 1975)) used a plaster of paris ring attached directly to the cornea and through mechanical linkages to recording pens. This technique evolved to the use of a modern contact lens to which a mounting stalk is attached. The contact lens is necessarily large, extending over the cornea and sclera (the lens is subject to slippage if the lens only covers the cornea). Various mechanical or optical devices have been placed on the stalk attached to the lens: reflecting phosphors, line diagrams, and wire coils have been the most popular implements in magneto-optical configurations. The principle method employs a wire coil, which is then measured moving through an electromagnetic field.[1] A picture of the search coil embedded in a scleral contact lens and the electromagnetic field frame are shown in Figure 5.2. The manner of insertion of the contact lens is shown in Figure 5.3. Although the scleral search coil is the most precise eye movement measurement method (accurate to about 5-10 arcseconds over a limited range of about 5°(Young & Sheena, 1975)), it is also the most intrusive method. Insertion of the lens requires care and practice. Wearing of the lens causes discomfort. This method also measures eye position relative to the head, and is not generally suitable for point of regard measurement.

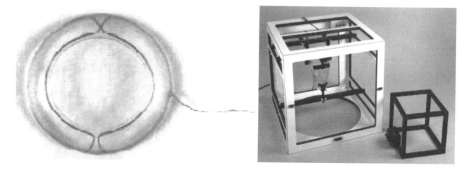

Fig. 5.2. Example of search coil embedded in contact lens and electromagnetic field frames for search coil eye movement measurement. Courtesy of Skalar Medical, PO Box 233, 2600 AE DELFT, The Netherlands <http://www.skalar.nl>. Reproduced with permission.

---

[1] This is similar in principle to magnetic position/orientation trackers often employed in Virtual Reality applications, e.g., Ascension's Flock Of Birds (FOB) uses this type of method for tracking the position/orientation of the head–see Chapter 7.

Fig. 5.3. Example of scleral suction ring insertion for search coil eye movement measurement. Courtesy of Skalar Medical, PO Box 233, 2600 AE DELFT, The Netherlands <http://www.skalar.nl>. Reproduced with permission.

5.3 Photo-Oculography (POG) or Video-Oculography (VOG)

This category groups together a wide variety of eye movement recording techniques involving the measurement of distinguishable features of the eyes under rotation/translation, e.g., the apparent shape of the pupil, the position of the limbus (the iris-sclera boundary), and corneal reflections of a closely situated directed light source (often infra-red). While different in approach, these techniques are grouped here since they often do not provide point of regard measurement. Examples of apparatus and recorded images of the eye used in photo- or video-oculography and/or limbus tracking are shown in Figure 5.4. Measurement of ocular features provided by these measurement techniques may or may not be made automatically, and may involve visual inspection of recorded eye movements (typically recorded on video tape). Visual assessment performed manually, e.g., stepping through a videotape frame-by-frame, can be extremely tedious and prone to error, and limited to the temporal sampling rate of the video device.

Automatic limbus tracking often involves the use of photodiodes mounted on spectacle frames (see Figure 5.4(b) and (c)), and almost always involves the use of invisible (usually infra-red) illumination (see Figure 5.4(d)). Several of these methods require the head to be fixed, e.g., either by using a head/chin rest, or a bite bar (Young & Sheena, 1975).

5.4 Video-Based Combined Pupil/Corneal Reflection

While the above techniques are in general suitable for eye movement measurements, they do not often provide point of regard measurement. To provide this measurement, either the head must be fixed so that the eye's position relative to the head and point of regard coincide, or multiple ocular features must be measured in order to disambiguate head movement from eye rotation. Two

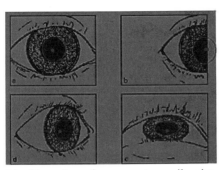

(a) Example of apparent pupil size. Courtesy of MetroVision, 4 rue des Platanes, 59840 Pérenchies, France <http://www.metrovision.fr>. Reproduced with permission.

(b) Example of infra-red limbus tracker apparatus. Courtesy of Applied Science Laboratories (ASL), 175 Middlesex Turnpike, Bedford, MA 01730 USA <http://www.a-s-l.com>. Reproduced with permission.

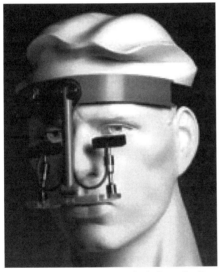

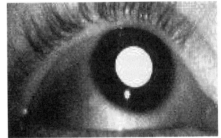

(c) Another example of infra-red limbus tracker apparatus, as worn by subject. Courtesy of Microguide, 1635 Plum Court, Downers Grove, IL USA <http://www.eyemove.com>. Reproduced with permission.

(d) Example of "bright pupil" (and corneal reflection) illuminated by infra-red light. Courtesy of LC Technologies, 9455 Silver King Court, Fairfax, VA USA <http://www.eyegaze.com>. Reproduced with permission.

Fig. 5.4. Examples of pupil, limbus, and corneal infra-red (IR) reflection eye movement measurements.

such features are the corneal reflection (of a light source, usually infra-red) and the pupil center (see Figure 5.4(d)).

Video-based trackers utilize relatively inexpensive cameras and image processing hardware to compute the point of regard in real-time. The apparatus may be table-mounted, as shown in Figure 5.5 or worn on the head, as shown in Figure 5.6. The optics of both table-mounted or head-mounted systems are essentially identical, with the exception of size. These devices, which are becoming increasingly available, are most suitable for use in interactive systems.

The corneal reflection of the light source (typically infra-red) is measured relative to the location of the pupil center. Corneal reflections are known as the Purkinje reflections, or Purkinje images (Crane, 1994). Due to the construction of the eye, four Purkinje reflections are formed, as shown in Figure 5.7. Video-based eye trackers typically locate the first Purkinje image. With appropriate calibration procedures, these eye trackers are capable of measuring a viewer's Point Of Regard (POR) on a suitably positioned (perpendicularly planar) surface on which calibration points are displayed.

Two points of reference on the eye are needed to separate eye movements from head movements. The positional difference between the pupil center and corneal reflection changes with pure eye rotation, but remains relatively constant with minor head movements. Approximate relative positions of pupil and first Purkinje reflections are graphically shown in Figure 5.8, as the left eye rotates to fixate 9 correspondingly placed calibration points. The Purkinje reflection is shown as a small white circle in close proximity to the pupil, represented by a black circle. Since the infra-red light source is usually placed at some fixed position relative to the eye, the Purkinje image is relatively stable while the eyeball, and hence the pupil, rotates in its orbit. So-called generation-V eye trackers also measure the fourth Purkinje image (Crane & Steele, 1985). By measuring the first and fourth Purkinje reflections, these dual-Purkinje image (DPI) eye trackers separate translational and rotational eye movements. Both reflections move together through exactly the same distance upon eye translation but the images move through different distances (thus changing their separation) upon eye rotation. This device is shown in Figure 5.9. Unfortunately, although the DPI eye tracker is quite precise, head stabilization may be required.

(a) Operator.

(b) Subject.

Fig. 5.5. Example of table-mounted video-based eye tracker.

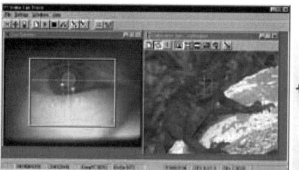

Fig. 5.6. Example of head-mounted video-based eye tracker. Courtesy of IOTA AB, EyeTrace Systems, Sundsvall Business & Tech. Center, S-851 71 Sundsvall, Sweden <http://www.iota.se>. Reproduced with permission.

5.5 Classifying Eye Trackers in "Mocap" Terminology

For readers familiar with motion capture ("mocap") techniques used in the special effects film industry, it is worthwhile to compare the various eye tracking methodologies with traditional mocap devices. Similarities between the two applications are intuitive and this is not surprising since the objective of both is recording the motion of objects in space. In eye tracking, the object measured is the eye, while in mocap, it is (usually) the joints of the body. Eye trackers can be grouped using the same classification employed to describe motion capture devices.

Electro-Oculography (EOG) is essentially an electro-mechanical device. In mocap applications, sensors may be placed on the skin or joints. In eye tracking, sensors are placed on the skin around the eye cavity. Eye trackers using a contact lens are effectively electro-magnetic trackers. The metallic stalk that is fixed to the contact lens is similar to the orthogonal coils of wire found in electro-magnetic sensors used to obtain the position and orientation of limbs and head in Virtual Reality. Photo-Oculography and Video-Oculography eye trackers are similar to the widely-used optical motion capture devices in special effects film, video, and game production. In both cases a camera is used to record raw motion, which is then processed by (usually) digital means to calculate the motion path of the object being tracked. Finally, video-based corneal reflection eye trackers are similar to optical motion capture devices which use reflective markers (worn by the actors). In both cases, an infra-red light source is usually used, for the reason that it is invisible to the human eye, and hence non-distracting.

PR, Purkinje reflections: 1, reflection from front surface of the cornea; 2, reflection from rear surface of the cornea; 3, reflection from front surface of the lens; 4, reflection from rear surface of the lens–almost the same size and formed in the same plane as the first Purkinje image, but due to change in index of refraction at rear of lens, intensity is less than 1% of that of the first Purkinje image; IL, incoming light; A, aqueous humor; C, cornea; S, sclera; V, vitreous humor; I, iris; L, lens; CR, center of rotation; EA, eye axis; $a \approx 6$mm; $b \approx 12.5$mm; $c \approx 13.5$mm; $d \approx 24$mm; $r \approx 7.8$mm (Crane, 1994).

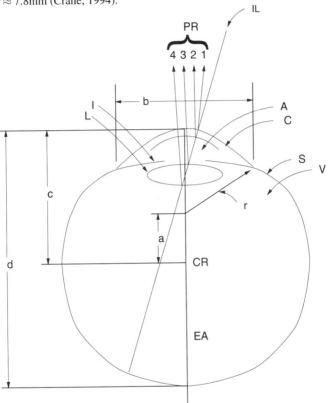

Fig. 5.7. Purkinje images. Reprinted in adapted form from Crane (1994) courtesy of Marcel Dekker, Inc. © 1994.

For a good introduction to motion capture for computer animation, as used in special effects film, video, and game production, see Menache (2000). Menache's book does a good job of describing motion capture techniques, although it is primarily aimed at practitioners in the field of special effects and animation production. Still, the description of mocap techniques, even at a comparatively large scale (i.e., capturing human motion vs. motion of the eye), provides a good classification scheme for eye tracking techniques.

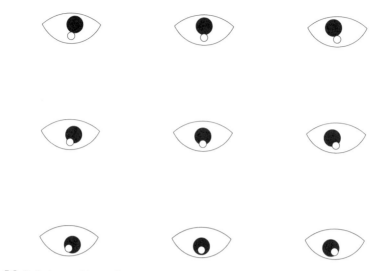

Fig. 5.8. Relative positions of pupil and first Purkinje images as seen by the eye tracker's camera.

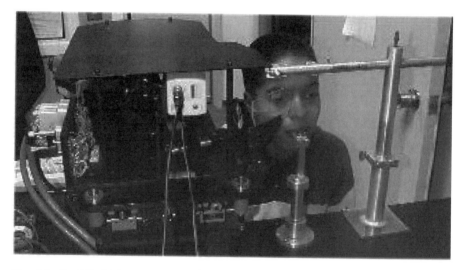

Fig. 5.9. Dual-Purkinje eye tracker. Courtesy of Fourward Optical Technologies, 2374 Hawthorne Ave., Buena Vista, VA 24416 USA <http://www.fourward.com>. Reproduced with permission.

5.6 Summary and Further Reading

For a short review of early eye tracking methods, see Robinson (1968, II). For another relatively more recent survey of eye tracking techniques, see Young and Sheena (1975). Although Young and Sheena's survey article is somewhat dated by today's standards, it is an excellent introductory article on eye movement recording techniques, and is still widely referenced. An up-to-date survey of eye tracking devices does not appear to be available, although the number of (video-based) eye tracking companies seems to be growing. Instead, one of the best comprehensive lists of eye tracking manufacturers is available the on the Internet, the Eye Movement Equipment Database (EMED) (see `<http://ibs.derby.ac.uk/emed/>`), maintained by David Wooding of the Institute of Behavioural Sciences, University of Derby, UK.

6. System Hardware Installation

Several types of eye trackers are available, ranging from scleral coils, EOG, to video-based corneal reflection eye trackers. While each has its benefits and drawbacks (e.g., accuracy vs. sampling rate), for graphical or interactive applications the video-based corneal reflection tracker is arguably the most practical device. These devices work by capturing video images of the eye (illuminated by an infra-red light source), processing the video frames (at video frame rates) and outputting the eye's x- and y-coordinates relative to the screen being viewed. The x- and y-coordinates are typically either stored by the eye tracker itself, or can be sent to the graphics host via serial cable. The advantage of the video-based eye tracker over other devices is that it is relatively non-invasive, fairly accurate (to about $1°$ visual angle over a $30°$ viewing range), and, for the most part, not difficult to integrate with the graphics system. The video-based tracker's chief limitation is its sampling frequency, typically limited by the video frame rate, 60Hz. Hence, one can usually expect to receive eye movement samples at least every 16ms (typically a greater latency should be expected since the eye tracker needs time to process each video frame, and the graphics host needs time to update its display).

6.1 Integration Issues and Requirements

Integration of the eye tracker into a graphics system depends chiefly on proper delivery of the graphics video stream to the eye tracker and the subsequent reception of the eye tracker's calculated 2D eye coordinates. In the following description of system setup, a complete graphics system is described featuring two eye trackers: one a table-mounted, monocular eye tracker set underneath a standard television display, the other a binocular eye tracker fitted inside a Head Mounted Display (HMD). Both displays are powered by the same graphics host. Such a laboratory has been set up within Clemson University's Virtual Reality Laboratory, and is shown in Figure 6.1. In Figure 6.1(b), the portion of the lab is shown where, on the right, the TV display is installed with the monocular table-mounted eye tracker positioned below, and on the left, the

HMD helmet is resting atop the d.c. electromagnetic head tracking units. The eye tracker unit (a PC) is situated between the HMD and the dual-head graphics display monitors to its right. Both HMD and TV (and graphics) displays are driven by the graphics host, which is out of view of the image (to the right and behind the camera). Figure 6.1(a) shows a close-up view of the eye tracker PC and the HMD. The four small TV monitors display the left and right scene images (what the user sees) and the left and right eye images (what the eye tracker sees).

The following system integration description is based on the particular hardware devices installed at Clemson's Virtual Reality Eye Tracking (VRET) laboratory, described here for reference. The primary rendering engine is a dual-rack, dual-pipe, SGI Onyx2® InfiniteReality2™ system with 8 raster managers and 8 MIPS® R12000™ processors, each with 8Mb secondary cache.[1] It is equipped with 8Gb of main memory and 0.5Gb of texture memory. The eye tracker is from ISCAN and the system includes both binocular cameras mounted within a Virtual Research V8 (high resolution) HMD as well as a monocular camera mounted on a remote pan-tilt unit. Both sets of optics function identically, capturing video images of the eye and sending the video signal to the eye tracking PC for processing. The pan-tilt unit coupled with the remote table-mounted camera/light assembly is used to non-invasively track the eye in real-time as the subject's head moves. This allows limited head movement, but a chin rest may optionally be used to restrict the position of the subject's head during experimentation to improve accuracy. The V8 HMD offers 640×480 resolution per eye with separate left and right eye feeds. HMD position and orientation tracking is provided by an Ascension 6 Degree-Of-Freedom (6DOF) Flock Of Birds (FOB), a d.c. electromagnetic system with a 10ms latency per sensor. A 6DOF tracked, hand-held mouse provides the user with directional motion control. The HMD is equipped with headphones for audio localization.

Although the following integration and installation guidelines are based on the equipment available at the Clemson VRET lab, the instructions should apply to practically any video-based corneal reflection eye tracker. Of primary importance to proper system integration are the following:

1. Knowledge of the video format the eye tracker requires as input (e.g., NTSC or VGA);
2. Knowledge of the data format the eye tracker generates as its output.

The first point is crucial to providing the proper image to the user as well as to the eye tracker. The eye tracker requires input of the scene signal so that it can

---

[1] Silicon Graphics, Onyx2, InfiniteReality, are registered trademarks of Silicon Graphics, Inc.

(a) HMD-mounted binocular eye tracker.

(b) Table-mounted monocular eye tracker.

Fig. 6.1. Virtual Reality Eye Tracking (VRET) lab at Clemson University.

overlay the calculated point of regard for display on the scene monitor which is viewed by the operator (experimenter). The second requirement is needed for proper interpretation of the point of regard by the host system. This is usually provided by the eye tracker manufacturer. The host system will need to be furnished with a device driver to read and interpret this information.

Secondary requirements for proper integration are the following:

1. The capability of the eye tracker to provide fine-grained cursor control over its calibration stimulus (a cross-hair or other symbol);
2. The capability of the eye tracker to transmit its operating mode to the host along with the eye x- and y point of regard information.

These two points are both related to proper alignment of the graphics display with the eye tracking scene image and calibration point. Essentially, both eye tracker capabilities are required to properly map between the graphics and eye tracking viewport coordinates. This alignment is carried out in two stages. First, the operator can use the eye tracker's own calibration stimulus to read out the extents of the eye tracker's displays. The eye tracker resolution may be specified over a 512×512 range, however, in practice it may be difficult to generate a graphics window which will exactly match the dimensions of the video display. Differences on the order of 1 or 2 pixels will ruin proper alignment. Therefore, it is good practice to first display a blank graphics window, then use the eye tracker cursor to measure the extents of the window in the eye tracker's reference frame. Since this measurement should be as precise as possible, fine cursor control is needed. Second, the eye tracker operates in three primary modes: reset (inactive), calibration, and run. The graphics program must synchronize with these modes, so that a proper display can be generated in each mode:

- Reset: display nothing (black screen) or a single calibration point.
- Calibration: display a simple small stimulus for calibration (e.g., a small dot or circle) at the position of the eye tracker's calibration stimulus.
- Run: display the stimulus required over which eye movements are to be recorded.

It is of the utmost importance that proper alignment is achieved between the eye tracker's calibration stimulus points and those of the graphics' display. Without this alignment, the data returned by the eye tracker will be meaningless.

Proper alignment of the graphics and eye tracking reference frames is achieved through a simple linear mapping between eye tracking and graphics window coordinates. This is discussed in Chapter 7, following notes on eye tracking system installation and wiring.

6.2 System Installation

The two primary installation considerations are wiring of the video and serial line cables between the graphics host and eye tracking systems. The connection of the serial cable is comparatively simple. Generation of the software driver for data interpretation is also straightforward, and is usually facilitated by the vendor's description of the data format and transmission rate. In contrast, what initially may pose the most problems is the connection of the video signal. It is imperative that the graphics host can generate a video signal in the format expected by both the graphics display (e.g., television set or HMD unit) and the eye tracker.

In the simplest case, if the host computer is capable of generating a video signal that is suitable for both the stimulus display and eye tracker, all that is then needed is a video splitter which will feed into both the stimulus display and eye tracker. For example, assume that the stimulus display is driven by an NTSC signal (e.g., a television), and the host computer is capable of generating a display signal in this format. (This is possible at the Clemson VRET lab since the SGI host can send a copy of whatever is on the graphics display via proper use of the `ircombine` command.) If the eye tracker can also be driven by an NTSC signal, then the installation is straightforward. If, however, the stimulus display is driven by VGA video (e.g., an HMD), but the eye tracker is driven by NTSC video, then the matter is somewhat more complicated. Consider the wiring diagram given in Figure 6.2. This schematic shows the dual display components, the HMD and TV, used for binocular eye tracking in a Virtual Reality environment, and monocular eye tracking over a 2D image or video display. The diagram features the necessary wiring of both left and right video channels from the host to the HMD and eye tracker, and a copy of the left video channel sent to the TV through the host's NTSC video output.

The HMD is driven by a horizontal-synch (h-sync) VGA signal. A switchbox is used (seen in Figure 6.1(a) just above the eye tracker keyboard) to switch the VGA display between the dual-head graphics monitors and the HMD. The HMD video control box diverts a copy of the left video channel through an active pass-through splitter back through the switchbox to the left graphics display monitor. The switchbox effectively "steals" the signal meant for the graphics displays and sends it to the HMD. The left switch on the switchbox

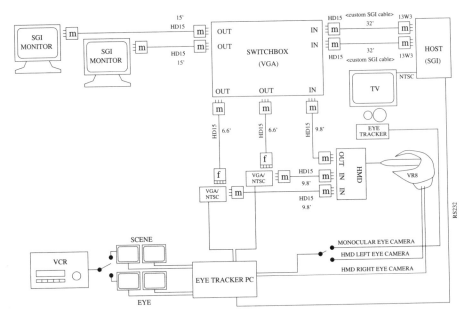

Fig. 6.2. Video signal wiring of the VRET lab at Clemson University.

has two settings: monitor or HMD. The right switch on on the switchbox has three settings: monitor, HMD, or both. If the right switch is set to monitor, no signal is sent to the HMD, effectively providing a biocular display in the HMD (instead of a binocular, or stereoscopic display). If the right switch is set to HMD, the graphics display blanks out since the HMD does not provide an active pass-through of the right video channel. If the right switch is set to both, the right video channel is simply split between the HMD and the monitor, re-sulting in a binocular display in both the HMD and on the monitors. This last setting provides no amplification of the signal and hence both the right LCD in the HMD and the right graphics monitor displays appear dim. This is mostly used for testing purposes.

The entire video circuit between the graphics host, the switchbox, and the HMD is VGA video. The eye tracker, however, operates on NTSC. This is the reason for the two VGA/NTSC converters which are inserted into the video path. These converters output an NTSC signal to the eye tracker and also pro-vide active pass-throughs for the VGA signal so that, when in operation, the VGA signal appears undisrupted. The eye tracker then processes the video sig-nals of the scene and outputs the signal to the scene monitors, with its own overlayed signal containing the calculated point of regard (represented by a cross-hair cursor). These two small displays show the operator what is in the user's field of view as well as as what s/he is looking at.

The eye images, in general, do not pose a complication since this video signal is exclusively processed by the eye tracker. In the case of the eye tracker at Clemson's VRET lab, both the binocular HMD eye cameras and the table-mounted monocular camera are NTSC and the signals feed directly into the eye tracker.

6.3 Lessons Learned from the Installation at Clemson

The eye tracker at Clemson contains hardware to process dual left and right eye and scene images. It can be switched to operate in monocular mode, for which it requires just the left eye and scene images. In this case, a simple video switch is used to switch the signal between the eye image generated by the left camera in the HMD and the camera on the table-mounted unit.

The SGI display monitors used at Clemson can be driven by either VGA video, or the default R, G, B video delivered by 13W3 video cables. To drive the HMD, VGA video was required, connected by HD15 cables. To connect the video devices properly, special 13W3-HD15 cables were needed. Although this seems trivial, custom-built cables were required. These cables are not cheap, and take a day or two to build and deliver. If timing and finances are a consideration, planning of the system down to the proper cabling is a must!

A problem that was particularly difficult to troubleshoot during the Clemson installation was the diagnosis of the format of the VGA signal emitted by the SGI host computer. Initially, before the eye tracker was installed, the HMD was tested for proper display. The output of the switchbox was used directly to drive the HMD. Everything functioned properly. However, after inserting the HMD into the video circuit, the eye tracker would not work. It was later found that the problem lay in the VGA/NTSC converters: these converters expect the more common VGA signal which uses a timing signal synchronized to the horizontal video field (h-sync signal; the horizontal and vertical sync signals are found on pins 13 and 14 of the VGA HD15 cable). The SGI host computer by default emits a sync-on-green VGA signal leaving pins 13 and 14 devoid of a timing signal. The VR HMD contains circuitry which will read either h-sync or sync-on-green VGA video and so functions quietly given either signal. The fault, as it turns out, was in an improper initial wiring of the switchbox. The switchbox was initially missing connections for pins 13 and 14 since these are not needed by a sync-on-green VGA signal. Once this was realized, the entire cable installation had to be disassembled and the switchbox had to be rewired. With the switchbox rewired, the custom 13W3 cables had to be inspected to confirm that these components carried signals over pins 13 and 14. Finally, a new display configuration had to be created (using SGI's `ircombine` command) to drive the entire circuit with the horizontal sync signal instead of the default sync-on-green. The moral here is: be sure of the video format required by all components, down to the specific signals transmitted on individual cable pins!

6.4 Summary and Further Reading

This chapter presented key points from an eye tracker's installation and its integration into a primarily computer graphics system. Although perhaps difficult to generalize from this particular experience (a case study if you will), nevertheless there are two points that are considered key to successful installation and usage of the device:

1. Signal routing and
2. Synchronization.

Signal routing refers to proper signal input and output, namely concerning video feeds (input to the eye tracker) and (typically rs232) serial data capture and interpretation (output from the eye tracker). Synchronization refers to the proper coordinate mapping of the eye tracker's reference frame to the application responsible for generating the visual stimulus that will be seen by the observer. Proper mapping will ensure correct registration of both eye tracker calibration and subsequent data analysis. The actual data mapping is carried out in software (see the next chapter), however, the hardware component that facilities proper software mapping is the eye tracker's capability of measuring the application's display extents in its own reference frame. This is usually provided by the eye tracker if it allows manual positioning and readout of its calibration cursor.

There are two primary sources where further information can be obtained on system installation and setup. First, of course, is the manufacturer's manual that will typically be included with the eye tracking device. Second, the best resource for installation and usage of the equipment is from the users themselves. Users of eye trackers typically report, in a somewhat formal way, what they use and how they use it in most technical papers on eye tracking research. These can be found in various journal articles, such as Vision Research, Behavior Research Methods, Instruments, and Computers (BRMIC), and conference proceedings. There are various conferences that deal with eye tracking, either directly or indirectly. For example, conferences that deal with computer graphics (e.g., SIGGRAPH, EuroGraphics, or Graphics Interface), human-computer interaction (e.g., SIGCHI), or Virtual Reality (e.g., VRST), may include papers that discuss the use of eye trackers, but their apparatus description may be somewhat indirect insofar as the objective of the report typically does not deal with eye tracker itself, rather it concentrates more on the results obtained through its application. Examples of such applications are given in Part III of this text. At this time, there are two main conferences that deal more directly with eye tracking: European Conference on Eye Movements (ECEM), and the US-based Eye Tracking Research & Applications (ETRA). Finally, the eye movement email listservs *eye-movements* and *eyemov-l* are excellent online "gathering places" of eye tracker researchers. Here, questions on eye trackers may be answered directly (and usually promptly) by users of similar equipment or even from other vendors.

7. System Software Development

In designing a graphical eye tracking application, the most important requirement is mapping of eye tracker coordinates to the appropriate application program's reference frame. The eye tracker calculates the viewer's Point Of Regard (POR) relative to the eye tracker's screen reference frame, e.g., a 512×512 pixel plane, perpendicular to the optical axis. That is, for each eye, the eye tracker returns a sample coordinate pair of x- and y-coordinates of the POR at each sampling cycle (e.g., once every ~ 16ms for a 60Hz device). This coordinate pair must be mapped to the extents of the application program's viewing window.

If a binocular eye tracker is used, two coordinate sample pairs are returned during each sampling cycle, x_l, y_l for the left POR and x_r, y_r for the right eye. A Virtual Reality (VR) application must, in the same update cycle, also map the coordinates of the head's position and orientation tracking device (e.g., typically a 6DOF tracker is used).

The following sections discuss the mapping of eye tracker screen coordinates to application program coordinates for both monocular and binocular applications. In the monocular case, a typical 2D image-viewing application is expected, and the coordinates are mapped to the 2D (orthogonal) viewport coordinates accordingly (the viewport coordinates are expected to match the dimensions of the image being displayed). In the binocular (VR) case, the eye tracker coordinates are mapped to the dimensions of the near viewing plane of the viewing frustum. For both calculations, a method of measuring the application's viewport dimensions in the eye tracker's reference frame is described, where the eye tracker's own (fine-resolution) cursor control is used to obtain the measurements. This information should be sufficient for most 2D image viewing applications.

For VR applications, subsequent sections describe the required mapping of the head position/orientation tracking device. While this section should generalize to most kinds of 6DOF tracking devices, the discussion in some cases is specific to the Ascension Flock Of Birds d.c. electromagnetic 6DOF tracker. This section discusses how to obtain the head-centric view and vectors from the matrix returned by the tracker, and also explains the transformation of an arbitrary vector using the obtained transformation matrix.[1] This latter derivation is used to transform the gaze vector to head-centric coordinates, which is initially obtained from, and relative to, the binocular eye tracker's left and right POR measurement.

Finally, a derivation of the calculation of the gaze vector in 3D is given, and a method is suggested for calculating the three-dimensional gaze point in VR.

7.1 Mapping Eye Tracker Screen Coordinates

When working with the eye tracker, the data obtained from the tracker must be mapped to a range appropriate to the given application. If working in VR, the 2D eye tracker data, expressed in eye tracker screen coordinates, must be mapped to the 2D dimensions of the near viewing frustum. If working with images, the 2D eye tracker data must be mapped to the 2D display image coordinates.

In general, if $x' \in [a, b]$ needs to be mapped to the range $[c, d]$, we have:

$$x = c + \frac{(x' - a)(d - c)}{(b - a)} \tag{7.1}$$

This is a linear mapping between two (one-dimensional) coordinate systems (or lines in this case). Equation (7.1) has a straightforward interpretation:

1. The value x' is translated (shifted) to its origin, by subtracting a.
2. The value $(x' - a)$ is then normalized by dividing through by the range $(b - a)$.
3. The normalized value $(x' - a)/(b - a)$ is then scaled to the new range $(d - c)$.
4. Finally, the new value is translated (shifted) to its proper relative location in the new range by adding c.

7.1.1 Mapping Screen Coordinates to the 3D Viewing Frustum

The 3D viewing frustum employed in the perspective viewing transformations is defined by the parameters `left`, `right`, `bottom`, `top`, `near`,

---

[1] Equivalently, the rotation quaternion be used, if this data is available from the head tracker.

far, e.g., as used in the OpenGL function call glFrustum(). Figure 7.1 shows the dimensions of the eye tracker screen (left) and the dimensions of the viewing frustum (right). Note that the eye tracker origin is the top-left of the screen while the viewing frustum's origin is bottom-left (this is a common discrepancy between imaging and graphics). To convert the eye tracker coordinates (x', y') to graphics coordinates (x, y), using Equation (7.1), we have:

$$x = \texttt{left} + \frac{x'(\texttt{right} - \texttt{left})}{512} \tag{7.2}$$

$$y = \texttt{bottom} + \frac{(512 - y')(\texttt{top} - \texttt{bottom})}{512} \tag{7.3}$$

Note that the term $(512 - y')$ in Equation (7.3) handles the y-coordinate mirror transformation so that the top-left origin of the eye tracker screen is converted to the bottom-left of the viewing frustum.

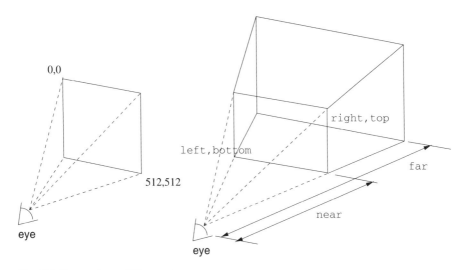

Fig. 7.1. Eye tracker to VR mapping.

If typical dimensions of the near plane of the graphics viewing frustum are 640×480, with the origin at $(0,0)$, then Equations (7.2) and (7.3) reduce to:

$$x = \frac{x'(640)}{512} = x'(1.3) \tag{7.4}$$

$$y = \frac{(512 - y')(480)}{512} = (512 - y')(.9375) \tag{7.5}$$

7.1.2 Mapping Screen Coordinates to the 2D Image

The linear mapping between eye tracker screen coordinates and 2D image coordinates is handled similarly. If the image dimensions are 640×480, then Equations (7.4) and (7.5) are used without change. Note that an image with dimensions 600×450 appears to fit the display of the TV in the Clemson VRET lab better.[2] In this case, Equations (7.4) and (7.5) become:

$$x = \frac{x'(600)}{512} = x'(1.171875) \qquad (7.6)$$

$$y = \frac{(512 - y')(450)}{512} = (512 - y')(.87890625) \qquad (7.7)$$

7.1.3 Measuring Eye Tracker Screen Coordinate Extents

The above coordinate mapping procedures assume that the eye tracker coordinates are in the range $[0, 511]$. In reality, the *usable*, or effective coordinates will be dependent on: (a) the size of the application window, and (b) the position of the application window. For example, if an image display program runs wherein an image is displayed in a 600×450 window, and this window is positioned arbitrarily on the graphics console, then the eye tracking coordinates of interest are restricted only to the area covered by the application window.

The following example illustrates the eye tracker/application program coordinate mapping problem using a schematic depicting the scene imaging display produced by an eye tracker manufactured by ISCAN. The ISCAN eye tracker is a fairly popular video-based corneal-reflection eye tracker which provides an image of the scene viewed by the subject and is seen by the operator on small ISCAN black and white scene monitors. Most eye trackers provide this kind of functionality. Since the application program window is not likely to cover the entire scene display (as seen in the scene monitors), only a restricted range of eye tracker coordinates will be of relevance to the eye tracking experiment. To use the ISCAN as an example, consider a 600×450 image display application. Once the window is positioned so that it fully covers the viewing display, it may appear slightly off-center on the ISCAN black/white monitor, as sketched in Figure 7.2.

The trick to the calculation of the proper mapping between eye tracker and application coordinates is the measurement of the application window's extents, in the eye tracker's reference frame. This is accomplished by using the

---

[2] A few of the pixels of the image do not fit on the TV display, possibly due to the NTSC flicker-free filter used to encode the SGI console video signal.

eye tracker itself. One of the eye tracker's most useful features, if available for this purpose, is its ability to move its crosshair cursor. The software may allow cursor fine (pixel by pixel) or coarse (in jumps of 10 pixels) movement. Furthermore, a data screen at the bottom of the scene monitor may indicate the coordinates of the cursor, relative to the eye tracker's reference frame.

To obtain the extents of the application window in the eye tracker's reference frame, simply move the cursor to the corners of the application window. Use these coordinates in the above mapping formulas. The following measurement "recipe" should provide an almost exact mapping, and only needs to be performed once. Assuming the application window's dimensions are fixed, the mapping obtained from this procedure can be hardcoded into the application. Here are the steps:

1. Position your display window so that it covers the display fully, e.g., in the case of the image stimulus, the window should just cover the entire viewing (e.g., TV or Virtual Reality Head Mounted Display (HMD)) display.
2. Use the eye tracker's cursor positioning utility to measure the viewing area's extents (toggle between course and fine cursor movement).
3. Calculate the mapping.
4. In Reset mode, position the eye tracker cursor in the middle of the display.

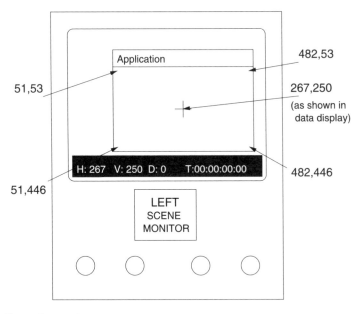

Fig. 7.2. Example mapping measurement.

An important consequence of the above is that following the mapping calculation, it should be possible to always position the application window in the same place, provided that the program displays the calibration point obtained from the eye tracker mapped to local image coordinates. When the application starts up (and the eye tracker is on and in Reset mode), simply position the application so that the central calibration points (displayed when the tracker is in Reset mode) line up.

To conclude this example, assume that if the ISCAN cursor is moved to the corners of the drawable application area, the measurements would appear as shown in Figure 7.2. Based on those measurements, the mapping is:

$$x = \frac{x' - 51}{(482 - 51 + 1)}(600) \tag{7.8}$$

$$y = 449 - \frac{y' - 53}{(446 - 53 + 1)}(450) \tag{7.9}$$

The central point on the ISCAN display is $(267, 250)$. Note that y is subtracted from 449 to take care of the image/graphics vertical origin flip.

7.2 Mapping Flock Of Birds Tracker Coordinates

In Virtual Reality, the position and orientation of the head is typically delivered by a real-time head tracker. In our case, we have a Flock Of Birds (FOB) d.c. electromagnetic tracker from Ascension. The tracker reports 6 Degree Of Freedom (6DOF) information regarding sensor position and orientation. The latter is given in terms of Euler angles. Euler angles determine the orientation of a vector in three-space by specifying the required rotations of the origin's coordinate axes. These angles are known by several names, but in essence each describes a particular rotation angle about one of the principal axes. Common names describing Euler angles are given in Table 7.1 and the geometry is sketched in Figure 7.3, where roll, pitch, and yaw angles are represented by R, E, and A, respectively. Each of these rotations is represented by the familiar homogeneous rotation matrices:

$$\text{Roll (rot z)} \quad = \quad \mathbf{R}_z = \begin{bmatrix} \cos R & \sin R & 0 & 0 \\ -\sin R & \cos R & 0 & 0 \\ 0 & 0 & 1 & 0 \\ 0 & 0 & 0 & 1 \end{bmatrix}$$

$$\text{Pitch (rot x)} \;=\; \mathbf{R}_x \;=\; \begin{bmatrix} 1 & 0 & 0 & 0 \\ 0 & \cos E & \sin E & 0 \\ 0 & -\sin E & \cos E & 0 \\ 0 & 0 & 0 & 1 \end{bmatrix}$$

$$\text{Yaw (rot y)} \;=\; \mathbf{R}_y \;=\; \begin{bmatrix} \cos A & 0 & -\sin A & 0 \\ 0 & 1 & 0 & 0 \\ \sin A & 0 & \cos A & 0 \\ 0 & 0 & 0 & 1 \end{bmatrix}$$

The composite 4×4 matrix, containing all of the above transformations rolled into one, is:

$$\mathbf{M} \;=\; \mathbf{R}_z \mathbf{R}_x \mathbf{R}_y \;=$$

$$\begin{bmatrix} \cos R \cos A + \sin R \sin E \sin A & \sin R \cos E & -\cos R \sin A + \sin R \sin E \cos A & 0 \\ -\sin R \cos A + \cos R \sin E \sin A & \cos R \cos E & \sin R \sin A + \cos R \sin E \cos A & 0 \\ \cos E \sin A & -\sin E & \cos E \cos A & 0 \\ 0 & 0 & 0 & 1 \end{bmatrix}$$

The FOB delivers a similar 4×4 matrix, $\mathbf{F}$,

$$\begin{bmatrix} \cos E \cos A & \cos E \sin A & -\sin E & 0 \\ -\cos R \sin A + \sin R \sin E \cos A & \cos R \cos A + \sin R \sin E \sin A & \sin R \cos E & 0 \\ \sin R \sin A + \cos R \sin E \cos A & -\sin R \cos A + \cos R \sin E \sin A & \cos R \cos E & 0 \\ 0 & 0 & 0 & 1 \end{bmatrix}$$

where the matrix elements are slightly rearranged, such that

$$\mathbf{M}[i, j] = \mathbf{F}[(i+1)\%3][(j+1)\%3],$$

e.g., row 1 of matrix $\mathbf{M}$ is now row 2 in matrix $\mathbf{F}$, row 2 is now row 3, and row 3 is now row 1. Columns are interchanged similarly, where column 1 of matrix $\mathbf{M}$ is now column 2 in matrix $\mathbf{F}$, column 2 is now column 3, and column 3 is now column 1. This "off-by-one" shift present in the Bird matrix may be due to the non-C style indexing which starts at 1 instead of 0.

Table 7.1. Euler angles.

rot y	rot x	rot x
yaw	pitch	roll
azimuth	elevation	roll
longitude	latitude	roll

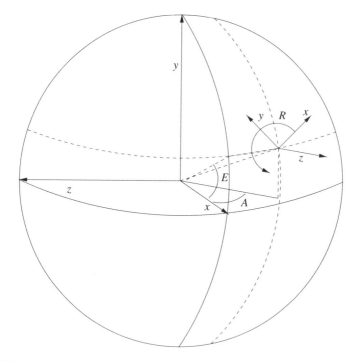

Fig. 7.3. Euler angles.

7.2.1 Obtaining the Transformed View Vector

To calculate the transformed view vector, $\mathbf{v}'$, assume initial view vector is $\mathbf{v} = (0,0,-1)$ (looking down z-axis), and apply the composite transformation (in homogeneous coordinates):[3]

$$\mathbf{vM} = \begin{bmatrix} 0 & 0 & -1 & 1 \end{bmatrix} *$$

$$\begin{bmatrix} \cos R\cos A + \sin R\sin E\sin A & \sin R\cos E & -\cos R\sin A + \sin R\sin E\cos A & 0 \\ -\sin R\cos A + \cos R\sin E\sin A & \cos R\cos E & \sin R\sin A + \cos R\sin E\cos A & 0 \\ \cos E\sin A & -\sin E & \cos E\cos A & 0 \\ 0 & 0 & 0 & 1 \end{bmatrix}$$

$$= \begin{bmatrix} -\cos E\sin A & \sin E & \cos E\cos A & 1 \end{bmatrix}$$

which is simply the third row of $\mathbf{M}$, negated. By inspection of the Bird matrix $\mathbf{F}$, the view vector is simply obtained by selecting appropriate entries in the matrix, i.e.,

$$\mathbf{v}' = \begin{bmatrix} -\mathbf{F}[0,1] & -\mathbf{F}[0,2] & \mathbf{F}[0,0] & 1 \end{bmatrix} \qquad (7.10)$$

---

[3] Recall trigonometric identities: $\sin(-\theta) = -\sin(\theta)$ and $\cos(-\theta) = \cos(\theta)$.

7.2.2 Obtaining the Transformed Up Vector

The transformed up vector is obtained in a similar manner to the transformed view vector. To calculate the transformed up vector, $\mathbf{u}'$, assume the initial up vector is $\mathbf{u} = (0, 1, 0)$ (looking up y-axis), and apply the composite transformation (in homogeneous coordinates):

$$\mathbf{uM} = \begin{bmatrix} 0 & 1 & 0 & 1 \end{bmatrix} *$$

$$\begin{bmatrix} \cos R \cos A + \sin R \sin E \sin A & \sin R \cos E & -\cos R \sin A + \sin R \sin E \cos A & 0 \\ -\sin R \cos A + \cos R \sin E \sin A & \cos R \cos E & \sin R \sin A + \cos R \sin E \cos A & 0 \\ \cos E \sin A & -\sin E & \cos E \cos A & 0 \\ 0 & 0 & 0 & 1 \end{bmatrix}$$

$$= \begin{bmatrix} -\sin R \cos A + \cos R \sin E \sin A & \cos R \cos E & \sin R \sin A + \cos R \sin E \cos A & 1 \end{bmatrix}$$

By inspection of the Bird matrix $\mathbf{F}$, the transformed up vector is simply obtained by selecting appropriate entries in the matrix, i.e.,

$$\mathbf{u}' = \begin{bmatrix} \mathbf{F}[2,1] & \mathbf{F}[2,2] & \mathbf{F}[2,0] & 1 \end{bmatrix} \tag{7.11}$$

Because of our setup in the lab, the z-axis is on the opposite side of the Bird transmitter (behind the Bird emblem on the transmitter). For this reason, the z-component of the up vector is negated, i.e.,

$$\mathbf{u}' = \begin{bmatrix} \mathbf{F}[2,1] & \mathbf{F}[2,2] & -\mathbf{F}[2,0] & 1 \end{bmatrix} \tag{7.12}$$

Note that the negation of the z-component of the transformed view vector does not make a difference since the term is a product of cosines.

7.2.3 Transforming an Arbitrary Vector

To transform an arbitrary vector, an operation similar to the transformations of the up and view vectors is performed. To calculate the transformed arbitrary vector, $\mathbf{w} = \begin{bmatrix} x & y & z & 1 \end{bmatrix}$, apply the composite transformation by multiplying by the transformation matrix $\mathbf{M}$ (in homogeneous coordinates):

$$\mathbf{wM} = \begin{bmatrix} x & y & z & 1 \end{bmatrix} *$$

$$\begin{bmatrix} \cos R \cos A + \sin R \sin E \sin A & \sin R \cos E & -\cos R \sin A + \sin R \sin E \cos A & 0 \\ -\sin R \cos A + \cos R \sin E \sin A & \cos R \cos E & \sin R \sin A + \cos R \sin E \cos A & 0 \\ \cos E \sin A & -\sin E & \cos E \cos A & 0 \\ 0 & 0 & 0 & 1 \end{bmatrix}$$

which gives transformed vector $\mathbf{w}'$,

$$\mathbf{w}' = \begin{bmatrix} x\mathbf{M}[1,1] + y\mathbf{M}[2,1] + z\mathbf{M}[3,1] \\ x\mathbf{M}[1,2] + y\mathbf{M}[2,2] + z\mathbf{M}[3,2] \\ x\mathbf{M}[1,3] + y\mathbf{M}[2,3] + z\mathbf{M}[3,3] \\ 1 \end{bmatrix}^T$$

To use the Bird matrix, there is unfortunately no simple way to select the appropriate matrix elements to directly obtain $\mathbf{w}'$. Probably the best bet would be to undo the "off-by-one" shift present in the Bird matrix. On the other hand, hardcoding the solution may be the fastest method. This rather inelegant, but luckily localized, operation looks like this:

$$\mathbf{w}' = \begin{bmatrix} x\mathbf{F}[1,1] + y\mathbf{F}[2,1] + z\mathbf{F}[0,1] \\ x\mathbf{F}[1,2] + y\mathbf{F}[2,2] + z\mathbf{F}[0,2] \\ x\mathbf{F}[1,0] + y\mathbf{F}[2,0] + z\mathbf{F}[0,0] \\ 1 \end{bmatrix}^T \tag{7.13}$$

Furthermore, as in the up vector transformation, it appears that the negation of the z component may also be necessary. If so, the above equation will need to be rewritten as:

$$\mathbf{w}' = \begin{bmatrix} x\mathbf{F}[1,1] + y\mathbf{F}[2,1] + z\mathbf{F}[0,1] \\ x\mathbf{F}[1,2] + y\mathbf{F}[2,2] + z\mathbf{F}[0,2] \\ -(x\mathbf{F}[1,0] + y\mathbf{F}[2,0] + z\mathbf{F}[0,0]) \\ 1 \end{bmatrix}^T \tag{7.14}$$

7.3 3D Gaze Point Calculation

The calculation of the gaze point in three-space depends only on the relative positions of the two eyes in the horizontal axis. The parameters of interest here are the three-dimensional virtual coordinates of the gaze point, (x_g, y_g, z_g), which can be determined from traditional stereo geometry calculations. Figure 7.4 illustrates the basic binocular geometry. Helmet tracking determines helmet position and orthogonal directional and up vectors, which determine viewer-local coordinates shown in the diagram. The helmet position is the origin, (x_h, y_h, z_h), the helmet directional vector is the optical (viewer-local) z-axis, and the helmet up vector is the viewer-local y-axis.

Given instantaneous, eye tracked, viewer-local coordinates (*mapped from eye tracker screen coordinates to the near view plane coordinates*), (x_l, y_l) and (x_r, y_r) in the left and right view planes, at focal distance f along the viewer-local z-axis, we can determine viewer-local coordinates of the gaze point,

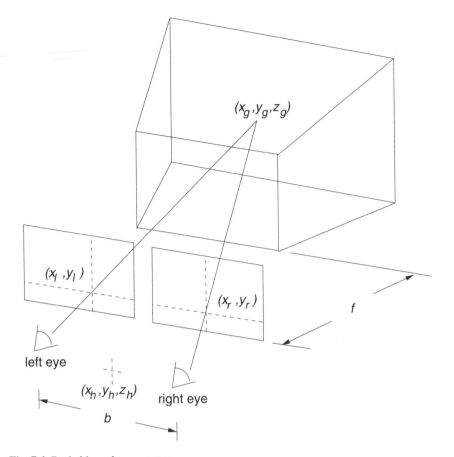

Fig. 7.4. Basic binocular geometry.

(x_g, y_g, z_g) by deriving the stereo equations parametrically. First, express both the left and right view lines in terms of the linear parameter s. These lines originate at the eye centers $(x_h - b/2, y_h, z_h)$ and $(x_h + b/2, y_h, z_h)$ and pass through (x_l, y_l, f) and (x_r, y_r, f), respectively. The left view line is (in vector form):

$$[(1-s)(x_h - b/2) + sx_l \quad (1-s)y_h + sy_l \quad (1-s)z_h + sf] \tag{7.15}$$

and the right view line is (in vector form):

$$[(1-s)(x_h + b/2) + sx_r \quad (1-s)y_h + sy_r \quad (1-s)z_h + sf] \tag{7.16}$$

where b is the disparity distance between the left and right eye centers, and f is the distance to the near viewing plane. To find the central view line originating at the local center (x_h, y_h, z_h), calculate the intersection of the left and right view lines by solving for s using the x-coordinates of both lines, as given in Equations (7.15) and (7.16):

$$(1-s)(x_h - b/2) + sx_l = (1-s)(x_h + b/2) + sx_r$$
$$(x_h - b/2) - s(x_h - b/2) + sx_l = (x_h + b/2) - s(x_h + b/2) + sx_r$$
$$s(x_l - x_r + b) = b$$

$$s = \frac{b}{x_l - x_r + b} \tag{7.17}$$

The interpolant s is then used in the parametric equation of the central view line, to give the gaze gaze point at the intersection of both view lines:

$$x_g = (1-s)x_h + s((x_l + x_r)/2)$$
$$y_g = (1-s)y_h + s((y_l + y_r)/2)$$
$$z_g = (1-s)z_h + sf$$

giving:

$$x_g = \left(1 - \frac{b}{x_l - x_r + b}\right)x_h + \left(\frac{b}{x_l - x_r + b}\right)\left(\frac{x_l + x_r}{2}\right) \tag{7.18}$$

$$y_g = \left(1 - \frac{b}{x_l - x_r + b}\right)y_h + \left(\frac{b}{x_l - x_r + b}\right)\left(\frac{y_l + y_r}{2}\right) \tag{7.19}$$

$$z_g = \left(1 - \frac{b}{x_l - x_r + b}\right)z_h + \left(\frac{b}{x_l - x_r + b}\right)f \tag{7.20}$$

Eye positions (at viewer local coordinates $(x_h - b/2, y_h, z_h)$ and $(x_h + b/2, y_h, z_h)$), the gaze point, and an up vector orthogonal to the plane of the three points then determine the view volume appropriate for display to each eye screen.

The gaze point, as defined above, is given by the addition of a scaled offset to the view vector originally defined by the helmet position and central view line in virtual world coordinates.[4] The gaze point can be expressed parametrically as a point on a ray with origin (x_h, y_h, z_h), the helmet position, with the ray emanating along a vector scaled by parameter s. That is, rewriting Equations (7.18), (7.19), and (7.20), we have:

$$x_g = x_h + s\left(\frac{x_l + x_r}{2} - x_h\right)$$
$$y_g = y_h + s\left(\frac{y_l + y_r}{2} - y_h\right)$$
$$z_g = z_h + s(f - z_h)$$

---

[4] Note that the vertical eye tracked coordinates y_l and y_r are expected to be equal (since gaze coordinates are assumed to be epipolar), the vertical coordinate of the central view vector defined by $(y_l + y_r)/2$ is somewhat extraneous; either y_l or y_r would do for the calculation of the gaze vector. However, since eye tracker data is also expected to be noisy, this averaging of the vertical coordinates enforces the epipolar assumption.

or, in vector notation,

$$\mathbf{g} = \mathbf{h} + s\mathbf{v} \tag{7.21}$$

where $\mathbf{h}$ is the head position, $\mathbf{v}$ is the central view vector and s is the scale parameter as defined in Equation (7.17). Note that the view vector used here is not related to the view vector given by the head tracker. It should be noted that the view vector $\mathbf{v}$ is obtained by subtracting the helmet position from the midpoint of the eye tracked x-coordinate and focal distance to the near view plane, i.e.,

$$\mathbf{v} = \begin{bmatrix} (x_l + x_r)/2 \\ (y_l + y_r)/2 \\ f \end{bmatrix} - \begin{bmatrix} x_h \\ y_h \\ z_h \end{bmatrix}$$
$$= \mathbf{m} - \mathbf{h}$$

where $\mathbf{m}$ denotes the left and right eye coordinate midpoint. To transform the vector $\mathbf{v}$ to the proper (instantaneous) head orientation, this vector should be normalized, then multiplied by the orientation matrix returned by the head tracker (see Section 7.2 in general and Section 7.2.3 in particular). This new vector, call it $\mathbf{m}'$, should be substituted for $\mathbf{m}$ above to define $\mathbf{v}$ for use in Equation (7.21), i.e.,

$$\mathbf{g} = \mathbf{h} + s(\mathbf{m}' - \mathbf{h}) \tag{7.22}$$

7.3.1 Parametric Ray Representation of Gaze Direction

Equation (7.22) gives the coordinates of the gaze point through a parametric representation (e.g., a point along a line) such that the depth of the three-dimensional Point Of Regard (POR) in world coordinates is valid only if $s > 0$. Given the gaze point, $\mathbf{g}$ and the location of the helmet, $\mathbf{h}$, we can obtain just the three-dimensional gaze vector, v which specifies the direction of gaze (but not the actual fixation point). This direction vector is given by:

$$v = \mathbf{g} - \mathbf{h} \tag{7.23}$$
$$= (\mathbf{h} + s\mathbf{v}) - \mathbf{h} \tag{7.24}$$
$$= s\mathbf{v} \tag{7.25}$$
$$= \left(\frac{b}{x_l - x_r + b} \right) \begin{bmatrix} (x_l + x_r)/2 - x_h \\ (y_l + y_r)/2 - y_h \\ (f - z_h) \end{bmatrix} \tag{7.26}$$

where v is defined as either $\mathbf{m} - \mathbf{h}$ as before, or as $\mathbf{m}' - \mathbf{h}$ as in Equation (7.22). Given the helmet position $\mathbf{h}$ and the gaze direction v, we can express the gaze direction via a parametric representation of a ray using a linear interpolant t:

$$\text{gaze}(t) = \mathbf{h} + t\mathbf{v}, \quad t > 0, \tag{7.27}$$

where h is the ray's origin (a point; the helmet position), and v is the ray direction (the gaze vector). (Note that adding h to $t\mathbf{v}$ results in the original expression of the gaze point, $\mathbf{g}$ given by Equation (7.21), provided $t = 1$.) The formulation of the gaze direction given by Equation (7.27) can then be used for testing virtual fixation coordinates via traditional ray/polygon intersection calculations commonly used in ray tracing.

7.4 Virtual Gaze Intersection Point Coordinates

In 3D eye tracking studies, we are often interested in knowing the location of one's gaze, or more importantly one's fixation, relative to some feature in the scene. In VR applications, we'd like to calculate the fixation location in the virtual world and thus identify the object of interest. The identification of the object of interest can be accomplished following traditional ray/polygon intersection calculations, as employed in ray tracing (Glassner, 1989).

The fixated object of interest is the one closest to the viewer which intersects the gaze ray. This object is found by testing all polygons in the scene for intersection with the gaze ray. The polygon closest to the viewer is then assumed to be the one fixated by the viewer (assuming all polygons in the scene are opaque).

7.4.1 Ray/Plane Intersection

The calculation of an intersection between a ray and all polygons in the scene is usually obtained via a parametric representation of the ray, e.g.,

$$\text{ray}(t) = r_o + t\mathbf{r_d} \tag{7.28}$$

where r_o defines the ray's origin (a point), and $\mathbf{r_d}$ defines the ray direction (a vector). Note the similarity between Equations (7.28) and (7.27)–there, h is the head position, and v is the gaze direction. To find the intersection of the ray with a polygon, calculate the interpolant value where the ray intersects each polygon, and examine all the intersections where $t > 0$. If $t < 0$, the object may intersect the ray, but behind the viewer.

Recall the plane equation $Ax + By + Cz + D = 0$, where $A^2 + B^2 + C^2 = 1$, i.e., A, B, and C define the plane normal. To obtain the ray/polygon intersection, substitute Equation (7.28) into the plane equation:

$$A(x_o + tx_d) + B(x_o + ty_d) + C(z_o + tz_d) + D = 0 \qquad (7.29)$$

and solve for t:

$$t = -\frac{Ax_o + Bx_o + Cz_o + D}{Ax_d + By_d + Cz_d} \qquad (7.30)$$

or, in vector notation:

$$t = \frac{-(\mathbf{N} \cdot r_o + D)}{\mathbf{N} \cdot \mathbf{r_d}} \qquad (7.31)$$

A few observations of the above simplify the implementation:

1. Here $\mathbf{N}$, the face normal, is really $-\mathbf{N}$, since what we're doing is calculating the angle between the ray and face normal. To get the angle, we need both ray and normal to be pointing in the same relative direction. This situation is depicted in Figure 7.5.
2. In Equation (7.31), the denominator will cause problems in the implementation should it evaluate to 0. However, if the denominator is 0, i.e., if $\mathbf{N} \cdot \mathbf{r_d} = 0$, then the cosine between the vectors is 0, which means that the angle angle between the two vectors is 90° which means the ray and plane are parallel and don't intersect. Thus, to avoid dividing by zero, and to speed up the computation, evaluate the denominator first. If it is sufficiently close to zero, don't evaluate the intersection further, we know the ray and polygon will not intersect.
3. Point 2. above can be further exploited by noting that if the dot product is greater than 0, then the surface is hidden to the viewer.

The first part of the intersection algorithm follows from the above and is given in Algorithm 7.1.

Algorithm 7.1 Ray/polygon intersection.

1: $v_d = \mathbf{N} \cdot \mathbf{r_d}$ // denominator
2: **if** $v_d < 0$ **then**
3:     $v_o = -(\mathbf{N} \cdot r_o + D)$ // numerator
4:     $t = v_o/v_d$
5: **end if**

In the algorithm, the intersection parameter t defines the point of intersection along the ray at the plane defined by the normal $\mathbf{N}$. That is, if $t > 0$, then the point of intersection, p, is given by:

$$p = r_o + t\mathbf{r_d} \qquad (7.32)$$

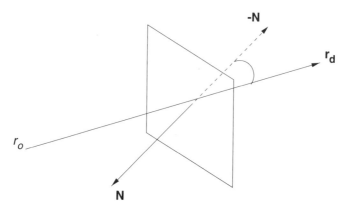

Fig. 7.5. Ray/plane geometry.

Note that the above only gives the intersection point of the ray and the (infinite!) plane defined by the polygon's face normal. Because the normal defines a plane of infinite extent, we need to test the point p to see if it lies within the confines of the polygonal region, which is defined by the polygon's edges. This is essentially the "point-in-polygon" problem.

7.4.2 Point-In-Polygon Problem

To test whether a point p lies inside a polygon (defined by its plane equation which specifies a plane of finite extent), we need to test the point against all edges of the polygon. To do this, the following algorithm is used:

For each edge:
1. Get the plane perpendicular to the face normal **N**, which passes through the edge's two vertices A an B. The perpendicular face normal, **N'**, is obtained by calculating the cross product of the original face normal with the edge, i.e.,

$$\mathbf{N'} = \mathbf{N} \times (B - A)$$

where the face vertices A and B are specified in counter-clockwise order. This is shown if Figure 7.6.

2. Get the perpendicular plane's equation by calculating D using either of A or B as the point on the plane (e.g., plug in either A or B into the plane equation, then solve for D).

3. Test point p to see if it lies "above" or "below" the perpendicular plane. This is done by plugging p into the perpendicular plane's plane equation and testing the result. If the result is greater than 0, then p is "above" the plane.

4. If p is "above" all perpendicular planes, as defined by successive pairs of vertices, then the point is "boxed in" by all the polygon's edges, and so must lie inside the polygon as originally defined by it face normal **N**.

Note that this is just one solution to the point-in-polygon problem—other, possibly faster, algorithms may be available.

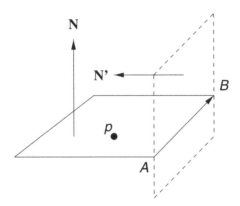

Fig. 7.6. Point-In-Polygon geometry.

The calculated intersection points p, termed here Gaze Intersection Points (GIPs) for brevity, are each found on the closest polygon to the viewer intersecting the gaze ray, assuming all polygons are opaque. The resulting ray casting algorithm generates a scanpath constrained to lie on polygonal regions within the virtual environment. Eye movement analysis is required to identify fixation points in the environment (see Chapter 9).

7.5 Data Representation and Storage

Data collection is straightforward. For 2D imaging applications, the point of regard can be recorded, along with the timestamp of each sample. Algorithm 7.2 shows a pseudocode sample data structure suitable for recording the point of regard. Because the number of samples may be rather large, depending on sampling rate and duration of viewing trial, a linked list may be used to dynamically record the data. In Algorithm 7.2, the data type `queue` denotes a linked list pointer.

Algorithm 7.2 2D imaging point of regard data structure.

1: **typedef struct** {	
2: queue *link*;	// *linked list node*
3: int *x,y*;	// *POR data*
4: double *t*;	// *timestamp (usually in ms)*
5: } PORnode;	

For 3D VR applications, the simplest data structure to record the calculated gaze point is similar to the 2D structure, with the addition of the third z component. Data captured, i.e., the three-dimensional gaze point, can then be displayed for review of the session statically in the VR environment in which it was recorded, as shown in Figure 7.7. In Figure 7.7 consecutive gaze points have been joined by straight line segments to display a three-dimensional scanpath. Note that in this visualization it may be easy to misinterpret the scanpath data with the position of the head. To disambiguate the two, head position (and tilt angle) may also be stored/displayed.

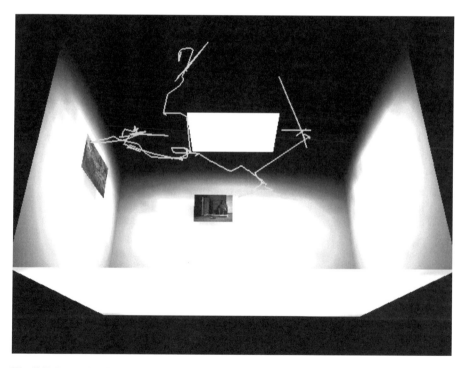

Fig. 7.7. Example of three-dimensional gaze point captured in VR. Courtesy of Tom Auchter and Jeremy Barron. Reproduced with permission, Clemson University.

There is one fairly important aspect of data storage which is worth mentioning, namely, proper labeling of the data. Because eye movement experiments tend to generate voluminous amounts of data, it is imperative to properly label data files. Files should encode the number of the subject viewing the imagery, the imagery itself, the trial number, and the state of the data (raw, or processed). One of the easiest methods for encoding this information is in the data filename itself (at least on file systems which allow long filenames).

7.6 Summary and Further Reading

This chapter focused on the main programming aspects of proper collection of raw eye movement data from the eye tracker, including appropriate mapping of raw eye movement coordinates. Coordinate mapping is crucial for both eye tracker calibration and subsequent eye movement coordination registration in the application reference frame.

Two important programming components have been omitted from this chapter, namely a mechanism for user interaction and for stimulus display. Generally, stimulus display requires knowledge of some sort of graphical Application Program Interface (API) which enables display of either digital images or graphics primitives (as required for VR). A particularly suitable API for both purposes is OpenGL (see <http://www.opengl.org>). For user interface development, particularly if developing on a Unix or Linux platform, the GTK Graphical User Interface (GUI) API is recommended. GTK is the freely available GNU Toolkit (see <http://www.gtk.org>).

8. System Calibration

Currently, most video-based eye trackers require calibration. This is usually a sequence of simple stimuli displayed sequentially at far extents of the viewing region. The eye tracker calculates the Point Of Regard (POR) by measuring the relative observed position of the pupil and corneal reflection at these locations, and then (most likely) interpolates the POR value at intermediate eye positions. The individual stimuli used for this purpose are simple white dots or cross-hairs on a black background. In cases where the eye tracker is used in an outside setting (e.g., for use during driving or while walking outside the lab), calibration marks may be made from simple targets such as tape or other visible markers fixed to objects in the environment. The purpose of calibration is to present a sequence of visible points at fairly extreme viewing angle ranges (e.g., upper-left, upper-right, lower-left, lower-right). These extrema points should be chosen to provide a sufficiently large enough coordinate range to allow the eye tracker to interpolate the viewer's POR between extrema points. Most (video-based) eye trackers provide built-in calibration techniques where a number of such extrema points (e.g., 3, 5, or 9 typically) are presented in order.

Apart from selection and/or presentation of properly distributed calibration stimulus points, a secondary but equally important goal of calibrating the eye tracker is correct adjustment of the eye tracker's optics and threshold levels to allow the device to properly recognize critical visual elements of the eyes. In the case of a video-based corneal reflection eye tracker, the device attempts to automatically detect the eye's pupil center and the center of the corneal reflection(s) (see Section 5.4). Corresponding to the these elements, the eye tracker software may offer a mechanism to adjust both pupil and corneal reflection detection thresholds. Both must be set so that the eye tracker can easily detect the centers of the pupil and the corneal reflection, without either losing detection of either or confusing these targets with distracting artifacts such as eyelashes, rims of glasses or contact lenses, to name a few usual problematic features. Figure 8.1 shows the eye images as seen by the eye tracker. Notice that the eye tracker has correctly labeled (with black crosshairs) the pupil center and one of two[1] corneal reflections while the subject was viewing each of five calibration targets. Figure 8.1(a) shows a fairly good eye image with properly identified pupil and corneal reflections; notice also the pupil threshold "bleeding" that is seen on the sclera to the bottom and left of the pupil. The pupil threshold may have been set a touch lower to eliminate this artifact.

Figures 8.1(d) and (e) show potential problems caused by long eye lashes. Eye lashes (particularly with heavy application of mascara), contact lenses, and eyeglasses may pose problems by providing reflecting surfaces in the scene which may interfere with the eye tracker's ability to locate the corneal reflection. However, eye tracking technology is improving–better image processing techniques, such as facial recognition, are helping to overcome these common calibration problems.

The general calibration procedure is composed of the following steps, to be performed by the operator after starting the system display/tracking program:

1. Move the application window to align it with the eye tracker's central calibration dot.
2. Adjust the eye tracker's pupil and corneal reflection threshold controls.
3. Calibrate the eye tracker.
4. Reset the eye tracker and run (program records the data).
5. Save recorded data.
6. Optionally calibrate again.

The final calibration step may be repeated to judge the amount of instrument slippage during the experimental trial (see Section 8.2 below). If the display/tracking program is one that relies on serial communication with the eye tracker, i.e., the program is responsible for generating the calibration stimulus points, the program must display the calibration stimulus at the same location the eye tracker is currently presenting to the viewer. This rather crucial application program requirement is discussed in the following section.

8.1 Software Implementation

If the application program is responsible for displaying the visual stimulus, i.e., the calibration dots and then later the test imagery, it is imperative that:

---

[1] The eye tracker used to generate the images of Figure 8.1 is a binocular eye tracker mounted inside a Head Mounted Display (HMD). Due to the proximity of the optics to each eye, two infra-red Light Emitting Diodes (LEDs) are used on either side of the eye. This is somewhat unusual; typically (for table-mounted, or remote eye trackers) one LED or other infra-red light source is used to project a single corneal reflection.

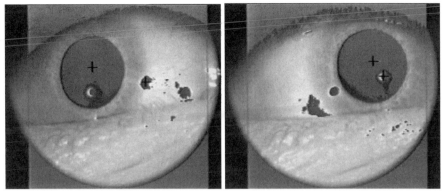

(a) Looking to upper-left.   (b) Looking to upper-right.

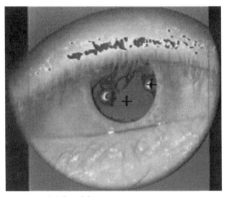

(c) Looking at screen center.

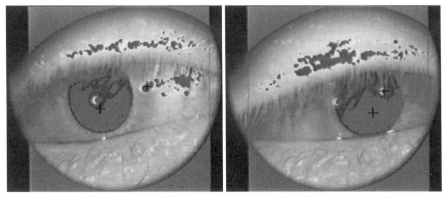

(d) Looking to bottom-left.   (e) Looking to bottom-right.

Fig. 8.1. Eye images during calibration (binocular eye tracking HMD).

(a) the program knows when to display the calibration dots or the test imagery, and (b) the calibration dots are aligned as precisely as possible so that the program's calibration dot is displayed at the location where the eye tracker expects it to be. The latter condition is satisfied partly by appropriate mapping of coordinates between application program window (viewport) coordinates and the eye tracker screen coordinates, and the initial placement of the application program window on the display screen (e.g., the window is moved to the appropriate position on the display device such as the TV or within the HMD). The former condition is met by obtaining a status word from the eye tracker itself. It is therefore imperative that the eye tracker provide this valuable piece of data along with the POR value(s).

Assuming the eye tracker provides its current state (along with the current eye coordinates x and y), the pseudocode for the usual graphics drawing routine, modified to draw the calibration points at the appropriate time, is shown in Algorithm 8.1. This routine is sensitive to the three possible modes of the eye tracker: RUN, RESET, and CALIBRATE. A typical double-buffered display update routine, for systems without eye trackers, would normally just be expected to set up the scene projection and object transformation matrices, then display the scene (stimulus) and swap buffers. In the modified routine, what is mostly needed is a way of drawing the calibration stimulus at the appropriate time.

Algorithm 8.1 Graphics draw/expose routine augmented with mode-sensitive eye tracking code.

```
 1: sceneProjection()                                  // projection (e.g., ortho or perspective)
 2: sceneMatrix()                                      // model (object) transformation(s)
 3: switch eye tracker state
 4:    case RUN:
 5:       if displayStimulus then
 6:          displayStimulus()                         // show image, VR, etc.
 7:       end if
 8:       break
 9:    case RESET:
10:    case CALIBRATE:
11:       drawCalibrationDot(x − 5, y − 5, x + 5, y + 5)      // show calibration dot
12:       break
13: end switch
14: swapbuffers()                                      // swap buffers
```

Notice that in the pseudocode of Algorithm 8.1 the calibration stimulus is displayed in both RESET and CALIBRATE states. This facilitates the initial alignment of the application window with the eye tracker's initial positioning of its cross-hairs. The default position is usually the center of the eye tracker screen. Provided the coordinate mapping routine is in place in the main loop of the program, the application program should correspondingly display its calibration dot at the center of its application window. The operator then simply positions the application window so that both points align in the eye tracker's scene monitor.

Notice also in Algorithm 8.1 that the stimulus scene (the image or VR environment) is only displayed if a display stimulus condition is satisfied. This condition may be set outside the drawing routine (e.g., in the main program loop) depending on some timing criterion. For example, if the experimental trial requires that an image be displayed for only 5 seconds, a timer in the main loop can be used to control the duration of the displayed stimulus by setting the display condition just after calibration has completed (state change from calibration to run). The timer then unsets the condition after the required duration has expired.

For VR applications, the draw routine may be preceded by viewport calls which determine which display, left or right, is drawn to. If the view is shifted horizontally, a binocular display is presented, otherwise, a biocular display is seen by the user. In VR, the calibration stimulus may require an orthographic projection call so that the 2D calibration points are not perturbed by a perspective transformation.

The main loop of the program is responsible for:

- reading the data from the eye tracker (and the head tracker if one is being used),
- mapping the eye tracker coordinates (and head tracker coordinates if one is being used),
- starting/stopping timers if the experimental conditions call for a specific stimulus duration period, and
- either storing or acting on the 2D (or 3D) gaze coordinates, if the application is diagnostic or gaze-contingent in nature, respectively.

This general algorithm is shown as pseudocode in Algorithm 8.2. In this instance of the main loop, a timer of length DURATION is used to control the duration of stimulus display.

A similar loop is needed for a VR application. There are a few additional requirements dealing with the head position/orientation tracker and calculation

---

Algorithm 8.2 Main loop (2D imaging application).

```
 1: while true do
 2:     getEyeTrackerData(x, y)                          // read serial port
 3:     mapEyeTrackerData(x, y)                          // map coordinates
 4:     switch eye tracker state
 5:         case RUN:
 6:             if !starting then
 7:                 starting = true
 8:                 startTimer()
 9:                 displayStimulus = true
10:                 redraw()                             // redraw scene event
11:             end if
12:             if checkTimer() > DURATION then
13:                 displayStimulus = false
14:                 redraw()                             // redraw scene event
15:             else
16:                 storeData(x, y)
17:             end if
18:             break
19:         case RESET:
20:         case CALIBRATE:
21:             starting = false
22:             redraw()                                 // redraw scene event
23:             break
24:     end switch
25: end while
```

---

of the three-dimensional gaze vector. The pseudocode for the main loop of a VR application is shown in Algorithm 8.3. The main loop routine for the VR application mainly differs in the calculation of the gaze vector, which is dependent on the stereoscopic geometry of the binocular system. The only other main difference is the data storage requirements. Instead of just the 2D point of regard, now the locations of both eyes (or the gaze vector) and/or the head need to be recorded. For gaze-contingent applications, instead of data storage, the gaze vector may be used directly to manipulate the scene or the objects within.

8.2 Ancillary Calibration Procedures

The calibration procedure described above is crucial for proper operation of the eye tracking device. Some devices will not even generate an eye movement data word until they have been calibrated. Device calibration is therefore necessary for any subsequent eye tracker device operation. Additional calibration routines may be used to test the device accuracy slippage before and after experimental trials or perhaps for calibration of subsidiary software measures, possibly internal to the application program. Denoting the device cali-

---

Algorithm 8.3 Main loop (3D Virtual Reality application).

```
 1: while true do
 2:     getHeadTrackerData(eye,dir,upv)                                    // read serial port
 3:     getEyeTrackerData(x_l,y_l,x_r,y_r)                                 // read serial port
 4:     mapEyeTrackerData(x_l,y_l,x_r,y_r)                                 // map coordinates
 5:     s = b/(x_l − x_r + b)               // calculate linear gaze interpolant parameter s
 6:     h = [eye_x,eye_y,eye_z]                                           // set head position
 7:     v = [(x_l + x_r)/2 − x_h,(y_l + y_r)/2 − y_h, f − z_h]    // calculate central view vector
 8:     transformVectorToHeadReferenceFrame(v)                    // multiply v by FOB matrix
 9:     g = h + sv                                                   // calculate gaze point
10:     switch eye tracker state
11:         case RUN:
12:             if !starting then
13:                 starting = true
14:                 startTimer()
15:                 displayStimulus = true
16:                 redraw()                                           // redraw scene event
17:             end if
18:             if checkTimer() > DURATION then
19:                 displayStimulus = false
20:                 redraw()                                           // redraw scene event
21:             else
22:                 storeData(x_l,y_l,x_r,y_r)
23:             end if
24:             break
25:         case RESET:
26:         case CALIBRATE:
27:             starting = false
28:             redraw()                                               // redraw scene event
29:             break
30:     end switch
31: end while
```

---

bration procedure as *external*, two ancillary procedures are presented, denoted as *internal*, since these procedures are more related to the application software rather than to the eye tracker itself. The first such ancillary internal calibration procedure is described for checking the accuracy slippage of the eye tracker. This optional procedure is described for the 2D eye movement recording program described above. The second ancillary internal calibration procedure is described for 3D software calibration, as used in VR when binocular (vergence) eye movement parameters are not measurable a priori.

8.2.1 Internal 2D Calibration

Each experimental trial includes three calibration steps. Calibration is divided into two procedures: *external* and *internal*. External calibration pertains to the proprietary eye tracker instrument calibration procedure specified by the manufacturer. Internal calibration pertains to the procedure developed within the developed application system for measurement of eye tracker accuracy. The eye tracker is externally calibrated using the vendor's proprietary 5- or 9-point calibration procedure. Internal calibration can be developed over any number of points, e.g., 30. The layout of the internal calibration points, denoted by the symbol + is shown in Figure 8.2. The point arrays are framed for clarity, horizontal and vertical lines do not appear during calibration. Figure 8.2 also shows the position of external calibration points overlayed on top of internal calibration points (in this case 9 points depicted by circles). Internal calibration is performed twice, immediately after external calibration (before stimulus display), and immediately after the stimulus display. Hence internal calibration is used to check the accuracy of the eye tracker before and after the stimulus display, to check for instrument slippage. The 30 internal calibration points can be shown in random order in a semi-interactive manner similar to the external calibration procedure. As each point is drawn on the screen (the subject is presented with a suitable simple stimulus, e.g., an × to minimize aliasing and flicker effects of an analog display), the application system waits for an input keypress before sampling eye movement data for a given sampling period (e.g., 800 ms). The input key delay allows the operator to observe eye stability on the eye tracker's eye monitor. The eye is judged to be stable once the eye tracker has repositioned the eye in the center of the camera frame. Recorded POR data is mapped to image coordinates in in real-time. Calibration data can be stored in a flat text file for later evaluation.

Internal calibration procedures provide the basis for two statistical measures: (1) the overall accuracy of the eye tracker, and (2) the amount of instrument slippage during stimulus viewing. The latter measurement gives an indication of the instrument accuracy during the viewing task, i.e., by recording loss of accuracy between the before- and after-viewing calibration procedures. A simple experiment illustrates these measures. An average of recorded POR data points (centroid; following coordinate mapping) is obtained and the error between the centroid and calibration point is calculated. Each two-dimensional Euclidean distance measurement is converted to the the full visual angle dependent on the viewing distance and calculated resolution of the television screen. A graphical example of this measurement is shown in Figure 8.3. The internal calibration locations are represented by +, sample measurements are represented by individual pixel dots, and centroid gaze positions are represented by circles, joined with corresponding calibration point by a line. The length of the line is the average error deviation in pixels.

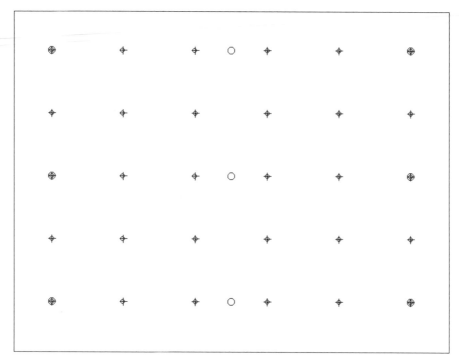

Fig. 8.2. Calibration stimulus—external (eye tracker) calibration points (circles) overlayed over internal calibration points (crosses).

To quantify overall eye tracker accuracy succinctly, the average calibration error can be obtained from each set of calibration points in order to calculate an overall average statistic for the eye tracker. The resulting average instrument error is an average statistic over all calibration runs.

To quantify instrument slippage, statistical analysis can be performed on before- and after-viewing mean error measure. Figure 8.4 shows a composite plot of before- and after- internal calibration stimulus display. Notice that, in this example, measured eye positions generally correspond well to calibration points. To quantify this correspondence, a one-way ANOVA can be performed on the means of the before- and after-viewing average error measures. If no significant difference is reported between the means, one can consider instrument slippage to be nominal.

8.2.2 Internal 3D Calibration

For 3D gaze point estimation, determination of the scalar s (dependent on inter-pupillary distance, or baseline, b) and focal distance f used in Equations

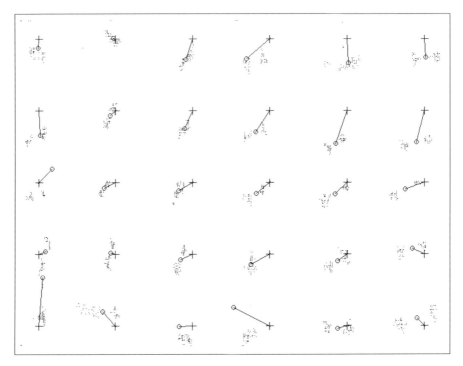

Fig. 8.3. Typical per-trial calibration data (1 subject) after stimulus viewing (avg. error: 1.77°).

(7.18)–(7.20) in Section 7.3 is difficult. Inter-pupillary distance cannot easily be measured in VR since the left and right eye tracking components function independently. That is, there is no common reference point. An internal 3D calibration procedure can be designed to estimate the inter-pupillary distance scaling factor s empirically (Duchowski, Medlin, Gramopadhye, Melloy, & Nair, 2001).

The 3D calibration relies on a specially marked environment, containing 9 clearly visible fixation targets, illustrated in Figure 8.5. Figure 8.5 shows scan-paths in a 3D Virtual Reality environment prior to eye movement analysis (see Chapter 9). Green spheres in the figure represent raw GIP data. The 9 × targets are distributed on 5 walls of the environment to allow head movement to be taken into account during analysis. Without a precise estimate of b and f, computed Gaze Intersection Points (GIPs) may appear stretched or compressed in the horizontal or vertical direction, as shown in Figure 8.5(a). Following manual manipulation of b and f values, GIP data is re-calculated and displayed interactively. The goal is to align the calculated GIP locations with the environmental targets which the user was instructed to fixate during calibration. An example of this type of adjustment is shown in Figure 8.5(b). Notice that the

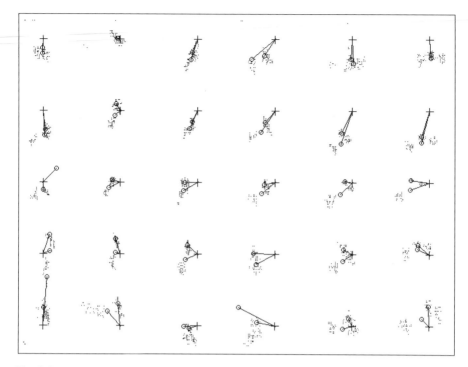

Fig. 8.4. Composite calibration data showing eye tracker slippage.

GIPs (represented by green, transparent spheres) are now better aligned over the red targets than the raw data in Figure 8.5(a). Once determined, appropriate scale factors are used to adjust each participant's eye movement data in all subsequent trials.

8.3 Summary and Further Reading

In summary, calibration of the eye tracker is essential toward proper eye movement data collection and analysis. The calibration procedure typically involves proper setting of the eye tracker's imaging optics (e.g., eye camera focus and contrast) and software threshold levels (e.g., pupil and corneal reflection). Although today's eye trackers have improved imaging and image processing components, traditional sources of calibration errors still exist and continue to pose problems. For example, contact lenses or eyeglasses may still interfere with some eye tracking devices; long eyelashes, heavy eye makeup, or "droopy" eyelids may prevent proper imaging of the eye, and in some scenarios the subject's own body and head movement may also cause imaging

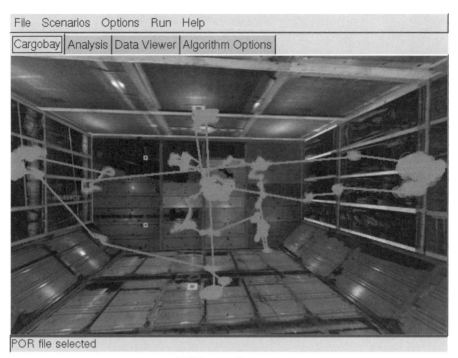

(a) Prior to adjustment.

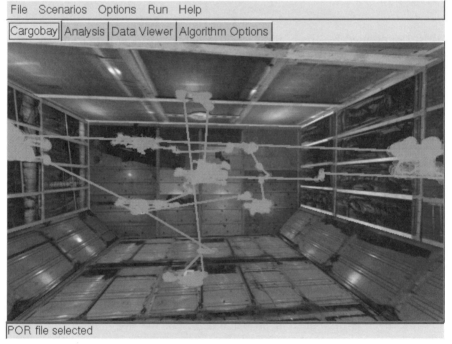

(b) Following adjustment.

Fig. 8.5. Adjustment of left and right eye scale factors.

problems. In some instances, head stabilization (e.g., the use of a chin rest) may be required.

Unfortunately, there is no widely accepted "how-to" text explaining proper calibration procedures. Most often, the eye tracker manufacturer's usage manual is the best primary source of information. Furthermore, by far the best advice for good manual calibration is experience and practice. Proper usage of the eye tracker, and in particular fast and "good" calibration, often comes with repeated use of the device. A current goal of eye tracking research is to devise a system for which calibration is not necessary, i.e., manufacturers are currently pursuing the development of an auto-calibrating eye tracker. While the current goal of doing away with calibration altogether has not yet been achieved, there is probably no better substitute for proper manual calibration than hands-on experience. It is often the case that new eye tracker operators are uncomfortable with the various controls of the device and may also feel nervous with directing human subjects to participate in eye tracking studies. However, with practice, one can quickly learn to properly use and calibrate an eye tracker. Once a sufficient comfort level is reached, proper calibration can be performed within a matter of seconds.

9. Eye Movement Analysis

The goal of eye movement measurement and analysis is to gain insight into the viewer's attentive behavior. As can be seen in Figure 8.5 at the end of Chapter 8, raw eye movement data, or perhaps data processed to a certain extent such as Gaze Intersection Point (GIP) data in Virtual Reality, may appear to be informative, however, without further analysis, raw data is for the most part meaningless. While intuitively (and from the knowledge of the task), it is possible to guess where the subject happened to be paying attention in the environment (over the internal calibration points, as s/he was instructed), it is not possible to make any further quantitative inferences about the eye movement data without further analysis. A method is needed to identify fixations—those eye movements which best indicate the locations of the viewer's (overt) visual attention.

As suggested in Chapter 4, eye movement signals can be approximated by linear filters. Fixations and pursuits can be modeled by a relatively simple neuronal feedback system. In the case of fixations, the neuronal control system is responsible for minimizing fixation error. For pursuit movements, the error is similarly measured as distance off the target, but in this case the target is non-stationary. Fixations and pursuits may be detected by a simple linear model based on linear summation.

The linear approach to eye movement modeling is an operational simplification of the underlying nonlinear natural processes (Carpenter, 1977). The linear model assumes that position and velocity is processed by the same neuronal mechanism. The visual system processes these quantities in different ways. The position of a target is signaled by the activation of specific retinal receptors. The velocity of the target, on the other hand, is registered by the firing rate (amplitude) of the firing receptors. Furthermore, nonlinearities are expected in most types of eye movements. Accelerational and decelerational considerations alone suggest the inadequacy of the linear assumption. Nevertheless, from a signal processing standpoint, linear filter analysis is sufficient for the localization of distinct features in eye movement signals. Although this approach is a poor estimate of the underlying system, it nonetheless establishes a useful approximation of the signal in the sense of pattern recognition.

The goal of eye movement signal analysis is to characterize the signal in terms of salient eye movements, i.e., saccades and fixations (and possibly smooth pursuits). Typically, the analysis task is to locate regions where the signal average changes abruptly indicating the end of a fixation and the onset of a saccade and then again assumes a stationary characteristic indicating the beginning of a new fixation. A hypothetical plot of an eye movement time course is shown in Figure 9.1. The graph shows the sought points in the signal where a saccade begins and ends. Essentially, saccades can be thought of as signal edges in time.

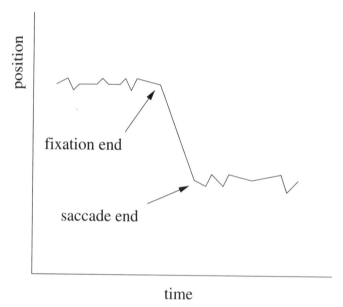

Fig. 9.1. Hypothetical eye movement signal.

Two main automatic types of approaches have been used to analyze eye movements: one based on summation (averaging), the other on differentiation.[1] In the first, the temporal signal is averaged over time. If little or no variance is found, the signal is deemed a candidate for fixation. Furthermore, the signal

---

[1] A third type of analysis, requiring manual intervention, relies on slowly displaying the time course of the signal (the scanpath), either in 1D or 2D, one sample at a time and judging which sample points lie outside the mean. This "direct inspection" method is rather tedious, but surprisingly effective.

is classified as a fixation, provided the duration of the stationary signal exceeds a predetermined threshold. This is the "dwell-time" method of fixation determination. In the second, assuming the eye movement signal is recorded at a uniform sampling rate, successive samples are subtracted to estimate eye movement velocity. The latter type of analysis is gaining favor, and appears more suitable for real-time detection of saccades. Fixations are either implicitly detected as the portion of the signal between saccades, or the portion of the signal where the velocity falls below a threshold.

Thresholds for both summation and differentiation methods are typically obtained from empirical measurements. The seminal work of Yarbus (1967) is often still referenced as the source of these measurements.

9.1 Signal Denoising

Before (or during) signal analysis, care must be taken to eliminate excessive noise in the eye movement signal. Inevitably, noise will be registered due to the inherent instability of the eye, and worse, due to blinks. The latter, considered to be a rather significant nuisance, generates a strong signal perturbation, which (luckily) may often be eliminated, depending on the performance characteristics of the available eye movement recoding device. It is often the case that either the device itself has capabilities for filtering out blinks, or that it simply returns a value of $(0,0)$ when the eye tracker "loses sight" of the salient features needed to record eye movements.

In practice, eye movement data falling outside a given rectangular range can be considered noise and eliminated. Using a rectangular region to denoise the (2D) signal also addresses another current limitation of eye tracking devices: their accuracy typically degrades in extreme peripheral regions. For this reason (as well as elimination of blinks), it may be sensible to simply ignore eye movement data falling outside the "effective operating range" of the device. This range will often be specified by the vendor in terms of visual angle. An example of signal denoising is shown in Figure 9.2, where an interior rectangular region 10 pixels within the image borders define the operating range. Samples falling outside this constrained interior image boundary were removed from the record (the original images measured 600×450 pixels).

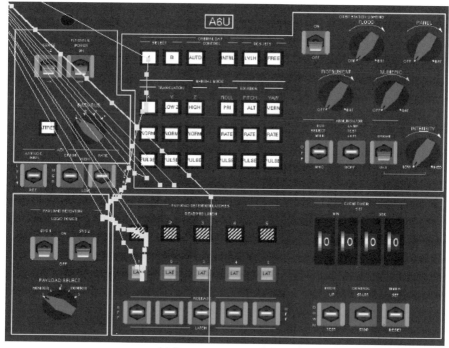

(a) Raw eye movement data.

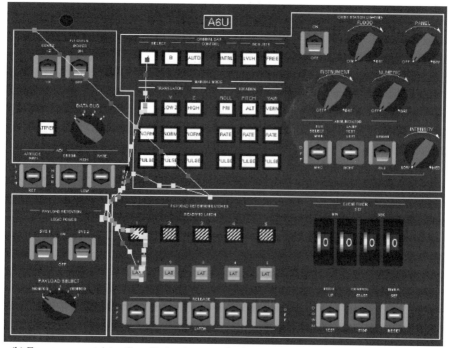

(b) Eye movement data after eliminating samples falling outside an interior region 10 pixels inside the image borders.

Fig. 9.2. Eye movement signal denoising. Courtesy of Wesley Hix, Becky Morley, and Jeffrey Valdez. Reproduced with permission, Clemson University.

9.2 Dwell-Time Fixation Detection

The dwell-time fixation detection algorithm depends on two characterization criteria:

1. Identification of a stationary signal (the fixation).
2. Size of time window specifying an acceptable range (and hence temporal threshold) for fixation duration.

An example of such an automatic saccade/fixation classification algorithm, suggested by Anliker (1976), determines whether M of N points lie within a certain distance D of the mean (μ) of the signal. This strategy is illustrated in Figure 9.3(a) where two N-sized windows are shown. In the second segment (positioned over a hypothetical saccade), the variance of the signal would exceed the threshold D indicating a rapid positional change, i.e., a saccade. The values of M, N, and D are determined empirically. Note that the value of N defines an a priori sliding window of sample times where the means and variances are computed. Anliker denotes this algorithm as the *position-variance method* since it based on the fact that a fixation is characterized by relative immobility (low position variance) whereas a saccade is distinguished by rapid change of position (high position variance).

9.3 Velocity-Based Saccade Detection

An alternative to the position-variance method is the *velocity detection method* (Anliker, 1976). In this velocity-based approach, the velocity of the signal is calculated within a sample window, and compared to a velocity threshold. If the sampled velocity is smaller than the given threshold, then the sample window is deemed to belong to a fixation signal, otherwise it is a saccade. The velocity threshold is specified empirically. Figure 9.3(b) shows the hypothetical eye movement time course with the sample window centered over the saccade with its velocity profile above.

Noting Yarbus' observation that saccadic velocity is nearly symmetrical (resembling a bell curve), a velocity-based prediction scheme can be implemented to approximate the arrival time and location of the next fixation. The next fixation location can be approximated as soon as the peak velocity is detected. Measuring elapsed time and distance traveled, and taking into account the direction of the saccade, the prediction scheme essentially mirrors the left half of the velocity profile (up to its peak) to calculate the saccade's end point.

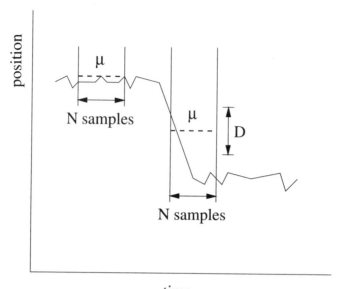

time
(a) Position-variance method.

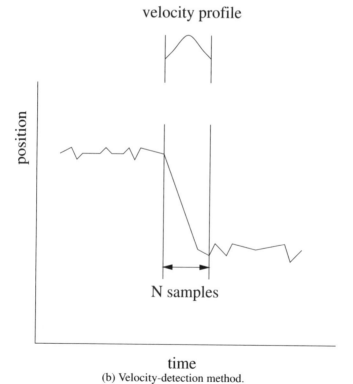

time
(b) Velocity-detection method.

Fig. 9.3. Saccade/fixation detection.

The position-variance and velocity-based algorithms give similar results, and both methods can be combined to bolster the analysis by checking for agreement. The velocity-based algorithm offers a slight advantage in that often short-term differential filters can be used to detect saccade onset, decreasing the effective sample window size. This speeds up calculation and is therefore more suitable for real-time applications. The image in Figure 9.2(b) was processed by examining the velocity,

$$v = \frac{\sqrt{(x_{t+1} - x_t)^2 + (y_{t+1} - y_t)^2}}{dt},$$

between successive sample pairs and normalizing against the maximum velocity found in the entire (short) time sequence. The normalized value, subtracted from unity, was then used to shade the sample points in the diagram. Slow moving points, or points approaching the speed of fixations are shown brighter than their faster counterparts.

Tole and Young (1981) suggest the use of short FIR (Finite Impulse Response) filters for saccade and fixation detection matching idealized saccade signal profiles. An idealized (discrete) saccade time course is shown in Figure 9.4(a). A sketch of the algorithm proposed by Tole and Young is presented in Figure 9.4(b), with corresponding FIR filters given in Figure 9.5. The algorithm relies on four conditions to detect a saccade:

$$|I_1| > A \tag{9.1}$$
$$|I_2| > B \tag{9.2}$$
$$Sgn(I_2) \neq Sgn(I_1) \tag{9.3}$$
$$T_{min} < I_2 - I_1 < T_{max} \tag{9.4}$$

If the measured acceleration exceeds a threshold (condition 9.1), the acceleration buffer is scanned forward to check for a second peak with opposite sign to that of I_1 (condition 9.3) and greater in magnitude than threshold B (condition 9.2). Amplitude thresholds are derived from theoretical values, e.g., corresponding to expected peak saccade velocities of 600°/s (visual angle). Condition (9.4) stipulates minimum and maximum durations for the forward search, also based on theoretical limits, e.g., saccade durations in the range 120ms–300ms. If all conditions are met, a saccade is detected.

Enhancements to the above algorithm, offered by Tole and Young, include adaptive threshold estimation, adjusted to compensate for recently observed signal noise. This is achieved by reformulating the thresholds A and B as functions of an RMS estimate of noise in the signal:

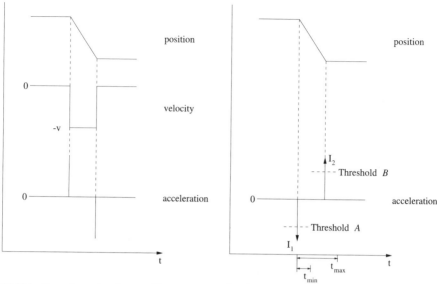

(a) Velocity and acceleration profiles.

(b) Finite Impulse Response (FIR) acceleration filter algorithm.

Fig. 9.4. Idealized saccade detection.

-1	-1	0	1	2	1	0	-1	-1

acceleration

-3	-2	-1	0	1	2	3

velocity

Fig. 9.5. Finite Impulse Response (FIR) filters for saccade detection.

$$\text{Threshold } A = 4000 \text{ deg/sec}^2 + \text{Accel/RMS} + \text{Accel/DC noise}$$
$$\text{Threshold } B = 4000 \text{ deg/sec}^2 + \text{Accel/RMS}.$$

The term

$$\text{Accel/DC noise} = \frac{2 \mid \Delta p \mid}{\Delta t^2}$$

estimates noise in acceleration, where Δt is the sampling interval and Δp is the peak-to-peak estimate of noise in position. This term can be estimated by measuring the peak-to-peak amplitude of fluctuations on the input signal when the average eye velocity over the last two seconds is less than 4 deg/sec. This estimate can be calculated once, for example, during calibration when the signal to noise ratio could be expected to remain constant over a test session, e.g., when the subject is known to be fixating a test target. This term may also be updated dynamically whenever velocity is low. The remaining estimate of noise in the output filter,

$$\text{Accel/RMS} = \sqrt{\frac{1}{T}\sum_{i=1}^{T}\left(\text{Accel}^2(i) - \overline{\text{Accel}}^{\,2}\right)},$$

assumes mean acceleration is zero for time windows greater than 4 seconds, i.e., $\overline{\text{Accel}} \to 0$ as $T > 4$s. The noise estimation term can be further simplified to avoid the square root,

$$\text{Accel/RMS} \leq \frac{1}{T}\sum_{i=0}^{T}\left|\,\text{Accel}(i)\,\right|, \quad T > 4\text{s}.$$

since $\sqrt{\sum \text{Accel}^2} \leq \sum |\,\text{Accel}\,|$. This adaptive saccade detection method is reported to respond well to a temporarily induced increase in noise level, e.g., when the subject is asked to grit their teeth.

9.4 Eye Movement Analysis in Three Dimensions

Analysis of eye movements in 3D, as recorded in Virtual Reality for example, follows the above algorithms, with slight modifications. In general, there are two important considerations: (a) elimination of noise and (b) identification of fixations. Noise may be present in the signal due to eye blinks, or other causes for the eye tracker's loss of proper imaging of the eye. These types of gross noise artifacts can generally be eliminated by knowing the device characteristics, e.g., the eye tracker outputs $(0,0)$ for eye blinks or other temporary missed readings. Currently there are two general methods for identification of fixations: the position-variance strategy, or the velocity-detection method. A third alternative may be a hybrid approach which compares the result of both algorithms for agreement.

The traditional two-dimensional eye movement analysis approach starts by measuring the visual angle of the object under inspection between a pair (or more) of raw eye movement data points in the time series (i.e., the POR data denoted by (x_i, y_i)). Given the distance between successive POR data points, $r = \|\,(x_i, y_i), (x_j, y_j)\,\|$, the visual angle, θ, is calculated by the equation: $\theta = 2\tan^{-1}(r/2D)$, where D is the (perpendicular) distance from the eyes to the viewing plane. Note that r and D, expressed in like units (e.g., pixels or inches), are dependent on the resolution of the screen on which the POR data was recorded. A conversion factor is usually required to convert one measure to the other (e.g., screen resolution in dots per inch (dpi) converting D to pixels). The visual angle θ and the difference in timestamps Δt between the POR data points allows velocity-based analysis, since $\theta/\Delta t$ gives eye movement velocity in degrees visual angle per second.

Note that the arctangent approach assumes that D is measured along the line of sight, which is assumed to be perpendicular to the viewing plane. Traditional 2D eye movement analysis methods can therefore be applied directly to raw POR data in the eye tracker reference frame. As a result, identified fixations could then be mapped to world coordinates to locate fixated ROIs within the virtual environment. A different approach may be followed by mapping raw POR data to world coordinates first, followed by eye movement analysis in three-space. The main difference of this approach is that calculated gaze points in three-space provide a composite three-dimensional representation of both the user's left and right eye movements. Applying the traditional 2D approach prior to mapping to (virtual) world coordinates suggests a component-wise analysis of left and right eye movements (in the eye tracker's reference frame) possibly ignoring depth. In three dimensions, depth information is implicitly taken into account prior to analysis. However, the assumption of a perpendicular visual target plane does not hold since the head is free to translate and rotate within 6 degrees of freedom.

Operating directly on Gaze Intersection Point (GIP) data in (virtual) world coordinates (see Chapter 7), a fixation detection algorithm based on estimation of velocity proceeds as follows. Given raw gaze intersection points in three dimensions, the velocity-based thresholding calculation is in principle identical to the traditional 2D approach, with the following important distinctions:

1. The head position, $\mathbf{h}$, must be recorded to facilitate the calculation of the visual angle.
2. Given two successive GIP data points in three-space, $\mathbf{p}_i = (x_i, y_i, z_i)$ and $\mathbf{p}_{i+1} = (x_{i+1}, y_{i+1}, z_{i+1})$, and the head position at each instance, $\mathbf{h}_i$ and $\mathbf{h}_{i+1}$, the estimate of instantaneous visual angle at each sample position, θ_i, is calculated from the dot product of the two gaze vectors defined by the difference of the gaze intersection points and averaged head position:

$$\theta_i = \cos^{-1} \frac{\mathbf{v}_i \cdot \mathbf{v}_{i+1}}{\| \mathbf{v}_i \| \, \| \mathbf{v}_{i+1} \|}, \quad i \in [0, n), \tag{9.5}$$

where n is the sample size and $\mathbf{v}_i = \mathbf{p}_i - \overline{\mathbf{h}}$ and $\overline{\mathbf{h}}$ is the averaged head position over the sample time period. Head position is averaged since the eyes can accelerate to reach a target fixation point much more quickly than the head (Watson, Walker, & Hodges, 1997).

With visual angle, θ_i, and timestamp difference between $\mathbf{p}_i$ and $\mathbf{p}_{i+1}$, the same velocity-based thresholding is used as in the traditional 2D case. No conversion between screen resolution and distance to target is necessary because all calculations are performed in world coordinates.

The algorithm generalizes to the use of wider filters (by changing the subscript $i+1$ to $i+k$ for $k > 1$) for improved smoothing. Using Equation (9.5) to calculate θ_i, only two successive data points are used to calculate eye movement velocity. This is analogous to the calculation of velocity using a 2-tap Finite Impulse Response (FIR) filter with coefficients $\{1, 1\}$.

To address excessive noise in the eye movement signal the 2-tap FIR filter can be replaced by a 5-tap FIR filter, as shown in Figure 9.7(a). Due to its longer sampling window, the 5-tap filter is more effective at signal smoothing (anti-aliasing). Following Tole and Young's (1981) work, an acceleration filter may also be used on 3D GIP data, with slight modification. The acceleration filter is shown in Figure 9.7(b), and is convolved with eye movement velocity data as obtained via either the 2-tap or 5-tap velocity filter. The filter responses resemble the real velocity and acceleration curves for a saccade characterized in Figure 9.6.

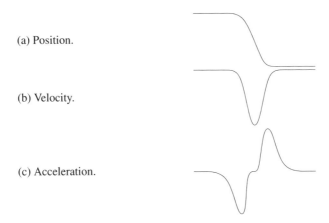

(a) Position.

(b) Velocity.

(c) Acceleration.

Fig. 9.6. Characteristic saccade signal and filter responses.

The 3D fixation eye movement analysis algorithm calculates the velocity and acceleration at each instantaneous estimate of visual angle, θ_i. Note that θ_i is

effectively a measure of instantaneous eye movement magnitude (i.e., amplitude), and therefore implicitly represents eye movement velocity. That is, the
signal resembles the positively oriented velocity peaks shown in Figure 9.6(b).
Withholding division by the time difference between successive samples (Δt)
facilitates the measurement of velocity with arbitrarily long filters.

Velocity is obtained via convolution with pattern-matching FIR filters of variable length. When convolved, these filters respond to sampled data with profiles matching that of the filter. These filters, denoted by $\mathbf{h}_k$, are essentially
unnormalized low-pass filters which tend to smooth and amplify the underlying signal. Division by the duration of the sampling window yields velocity,
i.e.,

$$\dot{\theta}_i = \frac{1}{\Delta t} \sum_{j=0}^{k} \theta_{i+j} \mathbf{h}_j, \quad i \in [0, n-k),$$

expressed in deg/s, where k is the filter length, $\Delta t = k - i$.

Acceleration is obtained via a subsequent convolution of velocity, $\dot{\theta}_i$, with the
acceleration filter, $\mathbf{g}_j$, shown in Figure 9.7(b). That is,

$$\ddot{\theta}_i = \frac{1}{\Delta t} \sum_{j=0}^{k} \dot{\theta}_{i+j} \mathbf{g}_j, \quad i \in [0, n-k),$$

where k is the filter length, $\Delta t = k - i$. The acceleration filter is essentially an
unnormalized high-pass differential filter. The resulting value, $\ddot{\theta}_i$, expressed in
deg/s^2, is checked against threshold A. If the absolute value of $\ddot{\theta}_i$ is greater than
A, then the corresponding gaze intersection point $\mathbf{p}_i$ is treated as the beginning
of a saccade. Scanning ahead in the convolved acceleration data, each subsequent point is tested in a similar fashion against threshold B to detect the end
of the saccade. Two additional conditions are evaluated to locate a saccade,
as given by Tole and Young. The four conditions are listed and illustrated in
Figure 9.8.

Note that these velocity and acceleration filters differ from those used by Tole
and Young. This is because Tole and Young applied their filters (the reverse of
these, essentially) to the positional eye movement signal ($\mathbf{p}$), while the filters
given here are applied to the signal amplitude (θ). Pseudocode of the technique
is presented in Algorithm 9.1.

9.4.1 Parameter Estimation

Thresholds are needed for saccade velocity, acceleration, and duration, since
the fixation detection algorithm relies on the detection of saccades. While al-

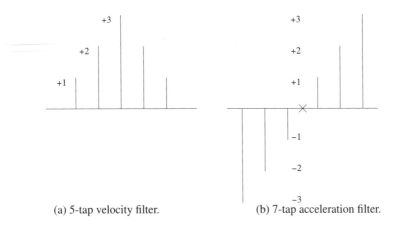

(a) 5-tap velocity filter.   (b) 7-tap acceleration filter.

Fig. 9.7. FIR filters.

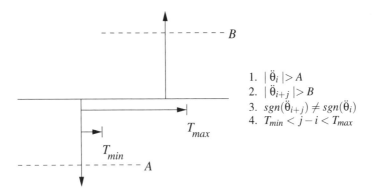

1. $|\ddot{\theta}_i| > A$
2. $|\ddot{\theta}_{i+j}| > B$
3. $sgn(\ddot{\theta}_{i+j}) \neq sgn(\ddot{\theta}_i)$
4. $T_{min} < j - i < T_{max}$

Fig. 9.8. Acceleration thresholding.

gorithm parameters may eventually be determined empirically, algorithm fine tuning is guided by a review of the literature, briefly summarized here for context.

The duration of saccades is related in a nonlinear manner to their amplitude over a thousandfold range (3′–50°) (Bahill, Clark, & Stark, 1975). Saccades of less than 15 or 20 degrees in magnitude are physiologically the most important since most naturally occurring saccades fall in this region. When looking at pictures, normal scanpaths are characterized by a number of saccades similar in amplitude to those exhibited during reading. The saccade "main sequence" describes the relationships between saccade duration, peak velocity, and magnitude (amplitude). Because saccades are generally stereotyped, the relationship between saccade amplitude and duration can be modeled by the linear equation $\Delta t = 2.2\theta + 21$ (Knox, 2001). Peak velocity reaches a soft saturation

---

Algorithm 9.1 Acceleration-based saccade detection.

Input: $\mathbf{p}(n)$, gaze intersection points, $\mathbf{h}(k)$, $\mathbf{g}(k)$, velocity and acceleration filters
Output: classification of each $\mathbf{p}_i$ as fixation or saccade

```
 1: for i = 0 to n − 1 do
 2:     θᵢ = cos⁻¹(vᵢ · vᵢ₊₁ / ‖ vᵢ ‖‖ vᵢ₊₁ ‖)              // calculate instantaneous visual angle
 3: end for
 4: for i = 0 to n − k − 1 do
 5:     θ̇ᵢ = θ̈ᵢ = 0                                          // initialize accumulation arrays (convolution results)
 6: end for
 7: for i = 0 to n − k − 1 do
 8:     for j = 0 to k do
 9:         θ̇ᵢ = θ̇ᵢ + θᵢ₊ⱼ * hⱼ                              // convolve with vel. filter
10:     end for
11: end for
12: for i = 0 to n − k − 1 do
13:     for j = 0 to k do
14:         θ̈ᵢ = θ̈ᵢ + θ̇ᵢ₊ⱼ * gⱼ                             // convolve with acc. filter
15:     end for
16: end for
17: for i = 0 to n − k − 1 do
18:     if | θ̈ᵢ | ≥ A (condition 1) then
19:         for j = i + Tₘᵢₙ to (n − k) − i ∧ (j − i) ≤ Tₘₐₓ (condition 4 implicit in loop) do
20:             if | θ̈ᵢ₊ⱼ | ≥ B ∧ sgn(θ̈ᵢ₊ⱼ) ≠ sgn(θ̈ᵢ) (conditions 2 & 3) then
21:                 for l = i to j do
22:                     pₗ = saccade
23:                 end for
24:             else
25:                     pᵢ = fixation
26:             end if
27:         end for
28:     end if
29: end for
```

---

limit up to about 15 or 20 degrees, but can range up to about 50°, reaching velocity saturation at about 1000 deg/s (Clark & Stark, 1975). In practice, the main sequence relationship between amplitude and velocity can be modeled by the asymptotic equation $\dot{\theta} = \lambda(1 - e^{-\theta/15})$, with velocity upper limit (asymptote λ) set to 750 deg/s (Hain, 1999).

According to accepted saccade amplitude estimates, measured instantaneous eye movement amplitude (θ) is expected to range up to about 20°. Example data captured in VR ranges up to 136°, with mean 1.5° (median 0.27°) and 9.7° s.d., which appears to be within normal limits, except for a few outliers (possibly due to head motion) (see Figure 9.9(a)).

For saccade detection via velocity filtering, a threshold of 130 deg/s may be chosen for both 2-tap and 5-tap filters. Using the asymptotic model of the main sequence relationship between saccade amplitude and velocity (limited by 750

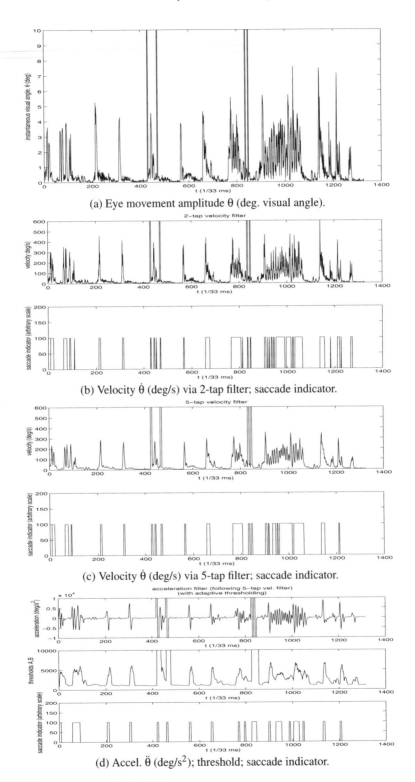

(a) Eye movement amplitude θ (deg. visual angle).

(b) Velocity θ̇ (deg/s) via 2-tap filter; saccade indicator.

(c) Velocity θ̇ (deg/s) via 5-tap filter; saccade indicator.

(d) Accel. θ̈ (deg/s²); threshold; saccade indicator.

Fig. 9.9. Eye movement signal and filter responses.

deg/s), this threshold should effectively detect saccades of amplitude roughly greater than $3°$. Observed velocity averages are reported in Table 9.1 (see also Figures 9.9(b) and 9.9(c)).

Saccade detection via acceleration filtering requires setting a larger number of parameters. Values of 10ms and 300ms for T_{min} and T_{max}, respectively, appear to cover a fairly wide range of saccade acceleration impulse pairs. The choice of the remaining threshold for saccade acceleration is made difficult since no applicable models of saccadic acceleration (e.g., a main sequence) could readily be found. In fact, unlike commonly listed limits of amplitude, duration, and velocity, there seems to be some disagreement regarding upper limits of acceleration. Peak acceleration has been reported to average at about 30,000 deg/s^2 in saccades of $10°$ with a saturation limit of 35,000 deg/s^2 for $\theta < 15°$, while other findings are given of $20°$ saccades with average peak acceleration of 26,000 deg/s^2 (Becker, 1989). Observed acceleration averages are reported in Tables 9.1 and 9.2 (see also Figure 9.9(d)). Following Tole and Young's acceleration filtering algorithm (incidentally, these authors report acceleration limits approaching 80,000 deg/s^2), the authors' recommended thresholds for saccade acceleration are suitable initial estimates (see below).

In Tole and Young's paper the authors point out variable noise characteristics dependent on the subject's actions (e.g., different noise profile while gritting teeth). To adapt to such signal changes the authors recommend an adaptive thresholding technique which dynamically adjusts the threshold, based on the current estimate of noise level. Indeed, a very large peak-to-peak acceleration signal variance may be observed (see Figure 9.9(d)). As the authors suggest, an adaptive thresholding technique may aid in automatically setting acceleration thresholds A and B:

$$A = B = 1,000 + \sqrt{\frac{1}{k}\sum_{i=0}^{k}(\ddot{\theta}_{i+k})^2}\ deg/s^2,$$

where k is the number of samples in time T proportional to the length of the acceleration filter, that is,

$$T = \frac{\text{filter length}}{\text{sampling rate}} = \frac{9}{30\text{Hz}} = 300\text{ms}.$$

This is a slightly different implementation of adaptive thresholding than Tole and Young's. The threshold value is slightly lower and its adaptive adjustment relies on explicit calculation of the acceleration Root Mean Squared (RMS). Also, the sampling window for this purpose is also much shorter from the

Table 9.1. Velocity algorithm comparisons.

	2-tap	5-tap
fixation groups	30	21
mean fixation duration (ms)	1079	1450
time spent in fixations	73%	69%
min $\dot{\theta}$ (deg/s)	0	.917
max $\dot{\theta}$ (deg/s)	12,385	5,592
avg. $\dot{\theta}$ (deg/s)	106	106
s.d. $\dot{\theta}$ (deg/s)	635	451

Table 9.2. Acceleration algorithm comparisons.

	2-tap		5-tap	
thresholding	adaptive	constant	adaptive	constant
fixation groups	20	17	17	14
mean fixation duration (ms)	1633	1583	1937	1983
time spent in fixations	74%	61%	74%	63%
min $\ddot{\theta}$ (deg/s^2)		-257,653		-182,037
max $\ddot{\theta}$ (deg/s^2)		248,265		167,144
avg. $\ddot{\theta}$ (deg/s^2)		4,453		3,966
s.d. $\ddot{\theta}$ (deg/s^2)		22,475		17,470

authors' recommended window of $T > 4$ sec. Finally, the adaptive technique give above employs a "look-ahead" scan of the acceleration data, suitable for off-line analysis. Changing the $i+k$ subscript to $i-k$ provides a "look-behind" scan which can be employed in real-time systems.

9.4.2 Fixation Grouping

The above algorithm classifies each GIP as either part of a fixation or saccade (see the saccade indicator plots in Figure 9.9). Once each GIP has been classified, each string of consecutive fixation GIPs is condensed to a single fixation point by finding the centroid of the group. However, due to the nature of the new algorithm, we observed that at times isolated noisy GIPs were also included in fixation groups. To prevent the inclusion of such outlying points a simple check can be implemented to verify that each fixation group's duration is greater than or equal to the minimum theoretical fixation duration (i.e., 150ms (Irwin, 1992)). Augmenting the acceleration-based saccade detection described above, this check can be considered as a position-variance component (emphasizing temporal coherence over spatial distribution) of a hybridized spatiotemporal approach to eye movement signal analysis.

9.4.3 Eye Movement Data Mirroring

Although the 3D eye movement analysis algorithm is mathematically robust at handling signal noise, the system is still susceptible to noise generated by the eye tracker. In particular, eye tracking equipment may randomly drop POR data. In some cases (e.g., during a blink), null POR values may be recorded for both left and right eyes. However, in some instances, only one eye's POR may be null while the other is not. This may occur due to calibration errors. To address this problem a heuristic technique may be used to mirror the non-null POR eye movement data (Duchowski et al, 2002). The table in Figure 9.10 shows an example of this technique. The left eye POR at time $t + 1$ is recorded as an invalid null point. To estimate a non-null left eye coordinate at $t + 1$, the difference between successive right eye POR values is calculated and used to update the left eye POR values at $t + 1$, as shown in the equation in Figure 9.10, giving $(x_{l_{t+1}}, y_{l_{t+1}}) = (-0.5 + dx, 0 + dy) = (-0.4, 0)$. Note that this solution assumes static vergence eye movements. It is assumed that the eyes remain at a fixed interocular distance during movement. That is, this heuristic strategy will clearly not account for vergence eye movements occurring within the short corrective time period.

Time	Left Eye	Right Eye
t	$(-0.5, 0)$	$(0.3, 0)$
$t+1$	$(0, 0)$	$(0.4, 0)$

$$dx = x_{r_{t+1}} - x_{r_t} = 0.4 - 0.3 = 0.1$$
$$dy = y_{r_{t+1}} - y_{r_t} = 0.0 - 0.0 = 0.0$$

Fig. 9.10. Heuristic mirroring example and calculation.

9.5 Summary and Further Reading

Eye movement analysis for 2D applications is fairly straightforward. The 3D eye movement analysis technique is resolution-independent and is carried out in three-space, which is particularly suitable for VR applications.

There are various collections of technical papers on eye movements, usually assembled from proceedings of focused symposia or conferences. A series of such books was produced by John Senders et al. in the 1970s and 1980s (see for example Monty and Senders (1976), Fisher et al (1981)). Good papers surveying eye movement analysis protocols have also recently appeared. Salvucci and Goldberg (2000) provide a good survey of current techniques (see also Salvucci and Anderson (2001)).

Part III

Eye Tracking Applications

10. Diversity and Types of Eye Tracking Applications

A wide variety of eye tracking applications exist, which can broadly be described within two categories, termed here as *diagnostic* or *interactive*. In its diagnostic role, the eye tracker provides objective and quantitative evidence of the user's visual and (overt) attentional processes. As an interface modality, the eye tracker serves as a powerful input device which can be utilized by a host of visually-mediated applications.

In general, in their diagnostic capacity, eye movements are simply recorded to ascertain the user's attentional patterns over a given stimulus. Diagnostic applications are distinguished by the unobtrusive use of the eye tracking device. In some cases, it may even be desirable to disguise the eye tracker so that potential subjects are not aware of its presence. Furthermore, the stimulus being displayed may not need to change or react to the viewer's gaze. In this scenario, the eye tracker is simply used to record eye movements for post-trial, off-line assessment of the viewer's gaze during the experiment. In this way, eye movement data may be used to objectively corroborate the viewer's point of regard, or overt locus of attention. For example, studies which test the appearance of some aspect of a display, say the location of an advertisement banner on a web page, may be bolstered by objective evidence of the user's gaze falling (or missing) the banner under consideration. Typical statistical measurements may include the number of fixations counted over the banner during a 5-minute "web browsing" session. Diagnostic eye tracking techniques are applicable (but not restricted) to the fields of Psychology (and Psychophysics), Marketing/Advertising, and Human Factors and Ergonomics.

Equipped with an eye tracker as an input device, an interactive system is expected to respond to, or interact with the user. Interactive applications are therefore expected to respond to the user's gaze in some manner. Such interactive systems may be classified by two application sub-types: *selective* and *gaze-contingent*. The latter can be further delineated in terms of display processing, as shown in the hierarchy in Figure 10.1. The archetypical interactive eye tracking application is one where the user's gaze is used as a pointing device.

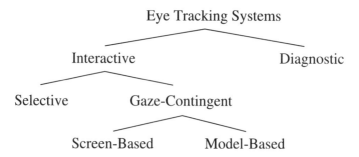

Fig. 10.1. Hierarchy of eye tracking applications.

This type of ocular interaction can be considered but one of a set of multimodal input strategies from the system's point of view (Hutchinson, 1993; Nielsen, 1993; Schroeder, 1993b, 1993a). An example of a system relying solely on gaze as input has been shown to be an important communication tool for quadriplegics, where the eyes are used for positioning a cursor over an oversized, projected keyboard. Using gaze to aid communication has also been explored in multi-party computer supported collaborative work systems (Vertegaal, 1999). Besides being used as a pointing device, knowledge of the user's gaze may be utilized to alter the display for speed-up purposes, as may be required in the rendering of complex virtual environments (McCormick, Batte, & Duchowski, 1996). Interactive eye tracking techniques are applicable (but not restricted) to the fields of Human-Computer Interaction, Visual Displays, and Computer Graphics.

10.1 Summary and Further Reading

A wide variety of (interdisciplinary) eye tracking applications have been and are currently being developed. The examples described and shown here by no means constitute a complete survey of the field, but it is hoped that they provide sufficient motivation to spur further interest in this fascinating area.

11. Neuroscience and Psychology

A wide assortment of eye tracking studies can be found in the increasingly related fields of Neuroscience and Psychology. Topics range from basic research in vision science to the investigation of visual exploration in aesthetics (e.g., perception of art). A useful approach to navigating through vast collections of early and contemporary literature is to (for the outset) dissociate high-level cognitive studies from those concerned with a low-level functional view of the brain. In this sense, to use a computational analogy, one can distinguish between the "hardware" (low-level brain circuitry) on which the "software" (high-level cognition) functions. In a complimentary view of the apparently disparate disciplines, neuroscience identifies the physiological components which are ultimately responsible for perception. In the context of vision and eye movements, knowledge of the physiological organization of the optic tract as well as of cognitive and behavioral aspects of vision is indispensable in obtaining a complete understanding of human vision.

To illustrate the inter-dependence of neuroscience and psychology, consider again the scene integration problem (see Chapter 1). Neurophysiological studies clearly identify the visual components involved in dynamic (or active) vision. That is, due to the limited informational capacity of the fovea, the eyes shift from point to point while scanning the visual field. The neuronal organization of retinal cells, which in a sense is the reason for eye movements, is well known. Furthermore, the general organization of foveo-peripheral vision has also been mapped along the magno- and parvo-cellular pathways leading to deeper regions of the brain and further still into regions implicated in higher cognitive functions. From psychological observations, we know that humans are aware of a large field of view, even though physiology does not permit a holistic, camera-like, capture of the entire scene in one exposure. This is the crux of the scene integration problem. Psychologists show us that we are fairly adept at maintaining a fairly accurate mental image of the visual scene in front of us. Indeed, various illusory pictures such as the Kanizsa (1976) square show us that we "see" more than what is physically there. The main question of how the brain is able to "piece together" small high-resolution snapshots of the scene remains a mystery.

While there are many eye-movement related topics examined in the fields of Neuroscience and Psychology, current neuroscientific trends related to the study of eye movements are briefly discussed touching on exemplary work investigating illusory contours, attention, and brain imaging. Focus is then shifted to psychological investigation of four applied perceptual examples: the study of eye movements in reading, during scene perception (including perception of art and film), visual search, and in natural tasks.

11.1 Neurophysiological Investigation of Illusory Contours

Neuroscientific investigation of vision and in the specific context of eye movements has been covered to a certain extent in previous chapters. Results of this research have led to the identification of various inter-connected components of vision, starting from retinal photoreceptors, and (more or less) ending in the cortical regions implicated in low-level vision. Important concepts of retinogeniculate pathways of vision as well as the important lower and higher-level visual brain regions have been well established. Augmenting traditional neuroscientific investigation with eye tracking devices can lead to further understanding of perceptually puzzling phenomena such as Kanizsa-type illusions.

Logothetis and Leopold (1995) conducted an experiment investigating the neural mechanisms underlying binocular rivalry–the alternating perceptions experienced when two dissimilar patterns are stereoscopically viewed. Single cells were recorded in visual areas V1, V2, and V4 while monkeys reported the perceived orientation of rivaling sinusoidal grating patterns. Monkeys were trained to perform a fixation and an orientation discrimination task. Both tasks required continuous fixation of a small central spot within a 0.8 × 0.8 degrees window. To confirm the location of fixations, eye movements were measured with a scleral coil technique. Logothetis and Leopold suggest that results of this study indicate that awareness of a visual pattern during binocular rivalry arises through interactions between neurons at different levels of visual pathways, and that the site of suppression is unlikely to correspond to a particular visual area, as often hypothesized on the basis of psychophysical observations. Together with earlier psychophysical evidence, the cell-types of modulating neurons and their overwhelming preponderance in higher rather than in early visual areas also suggests the possibility of a common neural mechanism underlying binocular rivalry as well as other bistable percepts, such as those experienced with ambiguous figures (e.g., Kanizsa-type illusions and the Necker cube). That is, perception of illusory contours may be physiologically related to stereoscopic perception.

11.2 Attentional Neuroscience

An important problem peculiar to eye tracking studies is that of the dissociation of visual attention from the point of regard. That is, most of the time, we can assume that one's visual attention is associated with the point of fixation. This is usually referred to as overt attention, since it is the component of visual attention (when associated with foveal vision) that can be measured overtly (i.e., by an eye tracker). However, it is certainly possible to fix one's gaze at a specific point, and yet move one's attention to a nearby region. Astronomers do this fairly regularly when looking for faint stars or star clusters with the naked eye. The little dipper is a good example of star clusters that is found more easily when one looks for it "off the fovea".

The dissociation of attention from ocular fixation poses a problem for eye tracking researchers. When examining a scanpath over a visual stimulus, we can often say that specific regions were "looked at", perhaps even fixated (following analysis of eye movements), however, we can not be fully confident that these specific regions were fully perceived. There is (currently) simply no way of telling what the brain was doing during a particular visual scan of the scene. Ideally, we would have to not only record the point of one's gaze, but also of one's brain activity. Research which combines eye tracking with traditional neuroscientific paradigms offers the dual benefit of monitoring brain activity as well as oculomotor function.

Investigating neuronal activity related to fixational eye movements, Snodderly, Kagan, and Gur (2001) used a double Purkinje image eye tracker (2–3 minarc resolution; 100 Hz sampling rate) to record the position of a macaque monkey's eye when fixating a light-emitting diode for 5 seconds. Action potentials were recorded from neurons in area V1 of three adult female macaque monkeys that were trained to hold visual fixation. Snodderly et al show that responses of V1 neurons to fixational eye movements are specific and diverse. Some cells are activated only by saccades, others discharge during drift periods, and most show a mixture of these two influences. Three types of eye movement activation were found: (1) "saccade-activated cells" discharged when a fixational saccade moved the activating region onto the stimulus, off the stimulus, and across the stimulus; (2) "position/drift cells" discharged during the intersaccadic (drift) intervals and were not activated by saccades that swept the activating region across the stimulus without remaining on it; (3) "mixed cells" fired bursts of activity immediately following saccades and continued to fire at a lower rate during intersaccadic intervals. The patterns of activity reflect the interactions between the stimulus, the receptive-field activating region, the temporal response characteristics of the neuron, and the retinal positions and image motions imparted by eye movements. The diversity of the activity patterns suggests that during natural viewing of a stationary scene some cortical neurons are carrying information about saccade occurrences and directions while other neurons are better suited to coding details of the retinal image.

Snodderly et al report that the two components of fixational eye movements, saccades and drifts, activate different subpopulations of V1 neurons in distinctive ways. Saccade-activated and mixed cells fire bursts of spikes when a fixational saccade moves the activating region onto the stimulus, off the stimulus, or across the stimulus. The sign of contrast (light or dark) is unimportant. These characteristics imply that the burst responses are conveying rather crude information about the details of the stimulus. For example, if an appropriately oriented stimulus contour activated the neuron following a saccade, it would be difficult to determine whether the contour remained in the activating region (saccade moved region onto contour) or was outside it (saccade moved region across contour). This ambiguity would have particular relevance when viewing complex natural images, because the neuronal discharge could be evoked either by a stimulus feature crossed by the activating region during the saccade, or by a completely different feature on which the region landed at the end of the saccade. Bursts of spikes fired by saccade and mixed cells following fixational saccades may suggest that the bursts are important sources of information about the visual scene. An alternative role for the saccade neurons might be to participate in the suppression of visual input associated with saccades. Snodderly et al argue that, theoretically, saccade neurons could participate in saccadic suppression by inhibiting other neurons that carry stimulus information, or by adding noise to the signal, thereby raising thresholds and making stimulus events undetectable at the times of saccades. This role would make the saccade neurons the source of saccadic suppression.

According to Snodderly et al, the position/drift neurons must play a quite different set of roles that are complementary to those of the other eye movement classes. Because the the position/drift neurons do not respond to saccades, they may be spared the potentially detrimental effects of saccade-related activity.

They signal accurately the position of stimulus features on the retina, and in many cases the sign of contrast. Thus, their activity could in principle be the basis for a reconstruction of the image.

Involved in attentional shifting behavior, the prefrontal (PF) cortex has long been thought to be central to the ability of choosing actions appropriate not only to the sensory information at hand but also according to the situation in which it is encountered (Asaad, Rainer, & Miller, 2000). Recent studies, reviewed by Asaad et al, indicate that the specific sensory, motor, and cognitive demands of the task (the behavioral context) can be an important factor in determining PF neural responses. For example, neural activity to an identical visual stimulus can vary as a function of which portion of that stimulus must be attended or with the particular motor response associated with it. Damage to the PF cortex of humans and monkeys tends to produce impairments when available sensory information does not clearly dictate what response is required. For example, PF lesions impair spatial delayed response tasks in which a cue is briefly flashed at one of two or more possible locations and the subject must direct an eye movement to its remembered location.

Asaad et al performed an eye tracked neurophysiological experiment to explore the role of the PF cortex. Subjects were two rhesus monkeys (with immobilized heads). One animal was implanted with an eye-coil to monitor eye movements, while an infra-red monitoring system from ISCAN was used for the second animal. Eye position was monitored at 100 Hz in both animals. Microsaccades and saccades were detected using a simple velocity threshold set at four times the standard deviation of the signal derived from the fixation period. Neural recording sites were localized using magnetic resonance imaging (MRI). Recording chambers were positioned stereotaxically over the left or right lateral prefrontal cortices of each animal, such that the principal sulcus and surrounding cortex, especially the ventrolateral PF cortex, was readily accessible. Recordings were made using arrays of eight durapuncturing, tungsten microelectrodes. Activity of up to 18 individual neurons could be recorded simultaneously in any given session. Signal from 210 neurons was recorded from the left lateral PF cortex of one monkey and 95 neurons from the right lateral PF cortex of the other. Monkeys performed an object memory task (delayed match-to-sample), an associative task (conditional visuomotor task), and a spatial memory task (spatial delayed response). The first two tasks shared common cue stimuli but differed in how these cues were used to guide behavior, whereas the latter two used different cues to instruct the same behavior. All three required the same motor responses. The associative task required the animals to associate a foveally presented cue stimulus with a saccade either to the right or left. The object task used the same cue stimulus as the associative task; however, in this case, they needed only to remember the identity of the cue and then saccade to the test object that matched it. Conversely the spatial task used small spots of light to explicitly cue a saccade to the right or left and required the monkeys to remember simply the response direction.

Asaad et al report that most of the 305 recorded neurons displayed a task-dependent change in overall activity, particularly in the fixation interval preceding cue presentation. Results show that for many PF neurons, activity was influenced by the task being performed. This influence included changes in their baseline firing rates, modulations of neuronal activity related to particular stimuli and behavioral responses, and difference in their firing rate profiles. Asaad et al suggest that results indicate the formal demands of behavior are represented within PF activity and thus support the hypothesis that one PF function is the acquisition and implementation of task context and the "rules" used to guide behavior.

11.3 Eye Movements and Brain Imaging

Recent investigations of eye movements and functional brain imaging simultaneously examine readings from an eye tracker and from a device which images brain activity. For example, Gamlin and Twieg (1997) have embarked on designing a combined visual display and eye tracking system for high-field fMRI (functional Magnetic Resonance Imaging) studies. The proposed project is a collaboration between a neuroscientist and biomedical engineer. The technological goal is the design, development, and implementation of a combined high resolution binocular display and eye tracking system for use in fMRI studies of the human brain. The scientific goal of the project is the study of neural control of saccadic eye movements, allowing investigation of stereopsis and depth perception, as well as permitting oculomotor studies of smooth pursuit, optokinetic nystagmus, vergence, and accommodative eye movements.

Systems which marry functional brain imaging with eye tracking can be used to at least corroborate a subject's fixation point while simultaneously recording cortical enhancement during attentional tasks. Presently, possibly due to prohibitive cost, combined eye tracking and brain imaging equipment is not widespread, although such devices are beginning to appear, e.g., see Figure 11.1.

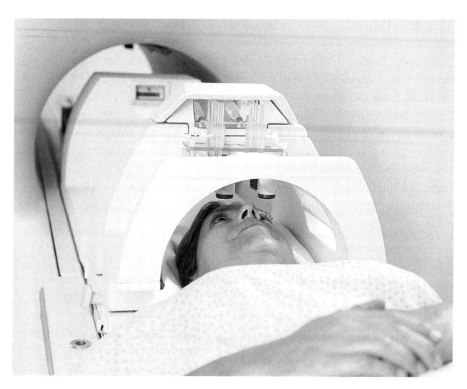

Fig. 11.1. Example of eye tracking fMRI scanner. Courtesy of SensoMotoric Instruments (SMI), 97 Chapel Street, Needham, MA 02492, USA <http://www.smiusa.com>. Reproduced with permission.

Employing a combined brain imaging and eye tracking device, possibly similar to the one shown in Figure 11.1, Özyurt, DeSouza, West, Rutschmann, and Greenlee (2001) used an fMRI device to compare the neural correlates of visually guided saccades in the step and gap paradigms. During task performance, saccadic eye movements were recorded with an MR-Eyetracker. Subjects viewed stimuli that were projected onto a screen attached to the front of the MR scanner. A block-design fMRI was used with alternating blocks of step, rest, and gap trials. Results from the study by Özyurt et al indicate significant task-related activity in striate and extrastriate cortex, the frontal eye fields, the supplementary motor area, parietal cortex, and angular gyrus, the frontal operculum, and the right prefrontal area 10. This type of research helps identify functional brain structures implicated in attentional behavior.

11.4 Reading

Perhaps the first well-known applied uses of eye trackers in the study of human (overt) visual attention were those conducted during reading experiments. An excellent book on eye movements was collected by Rayner (1992), an influential researcher in eye movements and reading. Rayner's collection of articles contains exemplary reading research and on scene perception. Rayner's (1998) recent article gives a rather comprehensive survey of eye tracking applications, reviewing studies of eye movements in reading and other information processing tasks such as music reading, typing, visual search, and scene perception. The major emphasis of the review is on reading as a specific example of cognitive processing.

Rayner (1998) reviews a good deal of previous work on eye movements synthesizing a large amount of information gleaned from over 100 years of research. While the reader is referred to Rayner's article for the complete review, three interesting examples of eye movement characteristics during reading are given here. First, eye movements differ somewhat when reading silently from reading aloud: mean fixation durations are longer when reading aloud or while listening to a voice reading the same text than in silent reading. Second, when reading English, eye fixations last about 200–250 ms and the mean saccade size is 7–9 letter spaces (see below). Third, eye movements are influenced by textual and typographical variables, e.g., as text becomes conceptually more difficult, fixation duration increases and saccade length decreases. Factors such as the quality of print, line length, and letter spacing influence eye movements. There is of course a good deal more that has been learned, however, here the methodology behind such discoveries is what is of primary interest. Below, three main experimental paradigms used in eye tracking experiments are discussed.

Three experimental paradigms, the *moving window*, *boundary*, and *foveal mask*, have been developed to explore eye movements. Although first developed for reading studies, these paradigms have since been adapted to other contexts such as scene perception (see below). In the moving window paradigm, or *gaze-contingent display change* paradigm, developed by McConkie and Rayner (1975), a window is sized to include a number of characters (e.g., 14) to the left and right of a fixated word. For example, the sentence

```
the quick brown fox jumped over the lazy dog
```

is presented as follows in 4 subsequent temporal instances (note the change of word `fox` to `cat`; the asterisks indicate fixation locations):

```
xhe quick brown xxxxxxxxxxxxxxxxxxxxxxxxxxxx
               *
xxxxxxxxk brown fox jumxxxxxxxxxxxxxxxxxxxxxx
              *
xxxxxxxxxxxxxxxxxcat jumped ovexxxxxxxxxxxxxxx
                  *
xxxxxxxxxxxxxxxxxxxxxxxxxxxxd over the laxxxxxx
                         *
```

The assumption with this technique is that when the window is as large as the region from which the reader can obtain information, there is no difference between reading in that situation and when there is no window. A related but reverse method, developed by Rayner and Bertera (1979) (see also Bertera and Rayner (2000)), places a foveal mask over a number of fixated characters (e.g., 7):

```
the qxxxxxxxown fox jumped over the lazy dog
    *
the quick brown cat jumpedxxxxxxxxhe lazy dog
                         *
```

This situation creates an artificial foveal scotoma and eye movement behavior if the situation is quite similar to the eye movement behavior of patients with real scotomas (Rayner, 1998).

In the boundary technique, developed by Rayner (1975), the stimulus changes as fixation crosses a predefined boundary. Rayner used eye movements to investigate reading since he found tachistoscopic methods inadequate: tachistoscopic (strobe-like) presentation of letters and words relies on the presentation of material for very brief exposures to exclude the possibility of an eye movement during the presentation. Prior to eye tracked reading studies, this method was often thought of as being analogous to a single fixation during reading. Based on his and others' research, Rayner argued that what subjects report from a tachistoscopic presentation cannot be taken as a complete specification of what they saw. The argument for eye movement recording over tachistoscopic displays carries over to scene perception and is discussed further in the next section.

In an example of the boundary technique below, a single critical target word is initially replaced by another word or by a non-word. The boundary paradigm allows the experimenter to be more diagnostic about what kind of information is acquired at different distances from fixation. In this technique, a word or nonword letter string is initially presented in a target location. However, when the reader's eye movement crosses an invisible boundary location, the initially presented stimulus is replaced by the target word. The amount of time that the subject looks at the target word is influenced by the relationship between the initially presented stimulus and the target word and the distance from the launch site to the target location. The fixation time on the target word thus allows the experimenter to make inferences about the type of information acquired from the target location when it was in parafoveal vision. For example, the word `fox` is changed to `cat` when the eyes cross the boundary.

```
the quick brown fox jumped over the lazy dog
            *       |
the quick brown cat jumped over the lazy dog
                |              *
```

The assumption is that if a reader obtains information from the initially presented stimulus, any inconsistency between what is available on the fixation after crossing the boundary and with what was processed on the prior fixation is registered in the fixation time on the target word.

Experiments using the above gaze-contingent techniques have shown, generally, (1) that readers typically acquire the visual information necessary for reading during the first 50–70 ms of a fixation, and (2) that a serial scan of letters in foveal vision does not occur. Furthermore, studies using these techniques consistently indicate that the size of the *perceptual span* is relatively small (e.g., for readers of alphabetical orthographies, such as English, Dutch, or French, the span extends from 3–4 letters to the left of fixation to about 14–15 letter spaces to the right of fixation), with a still smaller *word identification span* (Rayner, 1998).

Experiments in reading can lead to descriptions of individuals' reading strategies, or perhaps even to suggestions for improvement of one's reading strategy. Previous findings have shown that when an orienting visual response (eye movement) is made to a target pair consisting of two neighboring but separated elements, the first saccade lands in between the two elements. This was called the *global effect* by Findlay (1992).[1] In examining the global effect in reading, O'Regan (1992) suggests that an 'optimal' viewing position may exist, where the time it takes to recognize the word is minimal. Due to oculomotor constraints or scanning strategies, the eye does not always land at the optimal spot. O'Regan discusses a combination of two observed strategies and tactics

---

[1] The effect may also be referred to as the *center of gravity* effect.

for reading, as shown in Table 11.1. A double fixation involves a second fixation that is inversely proportional (in duration) to the first one. If far from optimal position, the fixation is very short (100-150ms). Single fixations may be long (up to 300ms). The optimal viewing position phenomenon is weak in reading, but strong in single word experiments. That is, it is not known what the 'special phenomenon' is in reading that accounts for the weakening of the optimal viewing position phenomenon. It may depend on reader's strategy–risky or careful. It is not known whether readers can choose to read faster or slower.

Table 11.1. Reading strategies/tactics.

Tactic	"Careful" strategy	"Risky" strategy
single fixation	one fixation in the critical region	one fixation in the critical region
double fixation	one fixation before critical region, one after (or the reverse–one fixation after and one backwards fixation before critical region)	two fixations in the critical region

The various types of gaze-contingent word display change techniques have had a major impact on reading research and scene perception because they allow the experimenter to control the information available to the subject at any point in time. Such experimental control over the nature of timing of the available information has enabled researchers to draw interesting conclusions about online aspects of reading and scene perception.

11.5 Scene Perception

Although certain reading patterns are easily recognized (e.g., left-to-right, top-to-bottom for English readers, or right-to-left for Hebrew), no apparent strategies for scene viewing have been easily discerned. Contrary to reading, there appears to be no canonical scanpath for particular objects (i.e., there is no particular "right way" to look at objects) (Kennedy, 1992). Kennedy suggests that the reading task is composed almost exclusively of saccades, while picture viewing is composed of shifts, pursuits, and drifts. There may be context differences at play. Continuing the debate about context effects for scenes and sentences, Kroll (1992) states that while there may be similarities between the two tasks, the tasks are very different. While eye movements in reading are to a large extent driven by the well-known, practiced task, we don't know what the 'glue' is that holds the scene together. Kroll states,

> One of the common problems in this research is to develop a set of tasks that will allow us to uniquely locate the interaction of context with object recognition over time.

Rayner (1998) recounts the traditionally held belief that examining the fine details of eye movements during scene perception is a high-cost, low-yield endeavor. Experiments using tachistoscopic presentations and eye movement recordings have lead to the conclusion that participants get the gist of a scene very early in the process of looking, sometimes even from a single brief exposure. Thus it has been advocated that the gist of the scene is abstracted on the first few fixations, and the remainder of the fixations on the scene are used to fill in details. Such findings give rise to the question of the value gained from information obtained from detailed eye movement analyses as people look at scenes. Rayner reviews several findings which support the contention that important conclusions about temporal aspects of scene perception can be obtained from eye movement analysis. He recounts the argument put forth by Loftus (1981) and Rayner and Pollatsek (1992) stating that tachistoscopic studies have not shown conclusively that they reveal a perceptual effect rather than the outcome of memory processes or guessing strategies.

Loftus (1981) presented results of a masked tachistoscopic study which suggest a model of picture encoding that incorporates the following propositions: (a) a normal fixation on a picture is designed to encode some feature of the picture; (b) the duration of a fixation is determined by the amount of time required to carry out the intended feature encoding; (c) the more features are encoded from a picture, the better the recognition memory will be from the picture. A major finding of Loftus' experiments was that with exposure time held constant, recognition performance increased with increasing numbers of fixations. When eye fixations are simulated tachistoscopically and their durations experimentally controlled, all traces of this phenomenon disappear. Moreover, Loftus' experiments suggest that within a fixation, visual information processing ceases fairly early, that is, acquired information reaches asymptote soon after the start of a fixation.

An eye fixation has the very salient property that it shifts the gaze to a new place in the picture. The problem identified with tachistoscopic exposures, which were meant to simulate fixations to new places, is that there is no guarantee that this occurred; it is entirely possible that subjects were simply holding their eyes steady throughout all tachistoscopic flashes. From an eye tracking experiment, Loftus draws the argument that given more places to look at in the picture, more information can be acquired from the picture. Additional (tachistoscopic) flashes are only useful insofar as they permit acquisition of information from additional portions of the picture. Information pertinent to subsequent recognition memory seems to be acquired only from the small 2° × 3° foveal region during a given fixation, and a fixation is useful only to the degree that it falls on a novel place in the picture.

The fixational *perceptual span* in scene perception mirrors that for reading with one important difference: meaningful information can be extracted much further from fixation in scenes than in text (Rayner, 1998). Objects located within about 2.6° from fixation are generally recognized, but recognition depends to some extent on the characteristics of the object. Qualitatively different information is acquired from the region 1.5° around fixation than from any region further from fixation. At high eccentricities, severely degraded information yields normal performance. This suggests that low-resolution information is processed in the more peripheral parts of the visual field, whereas high-resolution information is processed in foveal vision. High spatial frequency information is more useful in parafoveal and peripheral vision than low spatial frequency information.

Rayner and Pollatsek (1992) concede that much of the global information about the scene background or setting is extracted on the initial fixation. Some information about objects or details throughout the scene can be extracted far from fixation. However, if identification of an object is important, it is usually fixated. The work discussed in Rayner and Pollatsek's paper indicates that this foveal identification is aided significantly by the information extracted extrafoveally. Rayner and Pollatsek conclude that it is necessary to study eye movements to achieve a full understanding of scene perception. They argue that if the question of interest is how people process scenes in the real world, understanding the pattern of eye movements will be an important part of the answer.

Rayner (1998) summarizes a number of other findings and claims, some controversial, eventually asserting that given the existing data, there is fairly good evidence that information is abstracted throughout the time course of viewing a scene. Rayner concludes that whereas the gist of the scene is obtained early in viewing, useful information from the scene is obtained after the initial fixations.

According to Henderson and Hollingworth (1998), there are at least three important reasons to understand eye movements in scene viewing. First, eye movements are critical for the efficient and timely acquisition of visual information during complex visual-cognitive tasks, and the manner in which eye movements are controlled to service information acquisition is a critical question. Second, how we acquire, represent, and store information about the visual environment is a critical question in the study of perception and cognition. The study of eye movement patterns during scene viewing contributes to an understanding of how information in the visual environment is dynamically acquired and represented. Third, eye movement data provide an unobtrusive, online measure of visual and cognitive information processing. Henderson and Hollingworth list two important issues for understanding eye movement control during scene viewing: *where* the fixation position tends to be centered during scene viewing, and *how long* the fixation position tends to remain centered at a particular location in a scene.

Henderson and Hollingworth (1998) review past results indicating that the positions of fixations within a scene are non-random, with fixations clustering on informative scene regions. However, the specific effect of semantic informativeness beyond that of visual informativeness on fixation position is less clear. Several metrics can be used to evaluate the relative informativeness of scene regions: at a macro level analysis, *total time* that a region is fixated in the course of scene viewing (the sum of the durations of all fixations in that region)–this measure is correlated with the number of fixations in that region. At a micro level analysis, several commonly used measures include *first fixation duration* (the duration of the initial fixation in a region), *first pass gaze duration* (the sum of all fixations from first entry to first exit in a region), and *second pass gaze duration* (the sum of all fixations from second entry to second exit in a region). Generally, first pass gaze durations are longer for semantically informative (i.e., inconsistent) objects. Semantically informative objects also tend to draw longer second pass and total fixation durations. The influence of semantic informativeness on the duration of the very first fixation on an object is less clear. That is, scene context has an effect on eye movements: fixation time on an object that belongs in a scene is less than fixation time on an object that does not belong (Rayner, 1998). However, it is not clear whether the longer fixations on objects in violation of the scene reflect longer times to identify those objects or longer times to integrate them into a global representation of the scene (it could also reflect amusement of the absurdity of the violating object in the given context).

Henderson (1992) asks the question what is the influence of the contextual constraint provided by a predictive scene on the identification of its constituent objects? For example, does the context serve to facilitate identification procedures for objects consistent with that scene? Concretely, can a cow be identified more accurately and/or more quickly if it is viewed in a farm scene rather than in a kitchen scene? Henderson summarizes the predominant view of the relation between object and scene identification as the *schema hypothesis*. While there are several variations on the schema theme, several commonalities define the hypothesis. According to the scheme hypothesis, a memory representation of a prototypical scene is quickly activated during scene viewing, and is used to develop expectations about likely objects. These expectations then influence the object identification processes. For example, under the schema hypothesis, the identification of a cow in a farm scene involves (1) quickly recognizing that the scene is an exemplar of the category "farm scene", (2) accessing from memory the schema for a farm scene, (3) using the information stored with the schema to generate "cow" and other object candidates likely to be found in a farm scene (and possibly their canonical spatial relationships), and (4) using the knowledge that a cow is likely in such a scene to aid object identification processes when the cow is encountered. While the above description suggests a serial model, this need not be a central assumption of the schema hypothesis. Henderson cautions against the schema hypothesis and suggests the use of eye tracking to search for a better model of scene perception.

Henderson (1992) offers two criticisms of the eye movement paradigm. First, it is likely that global measures of fixation time, such as the total time spent on an object during the course of scene viewing, and the gaze duration on an object (the time of all initial fixations on an object prior to leaving that object for the first time) reflect post-identification processes. Thus, it is likely that gaze duration in scene processing reflects other processes beyond object identification. Henderson suggests that the preferred fixation measure is the true first fixation duration, or the duration of time from the initial landing of the eyes on an object until the eyes move to any other location, including another location on the object. Second, the basic premise of the eye movement paradigm is that the results will reflect normally occurring visual-cognitive processes because subjects can view scenes in a natural manner. However, unlike reading, where the overall task is arguably transparent, subjects must be given an orienting task when they view a scene. Unfortunately, viewing behavior and eye movement patterns change as a function of the viewing task given to the subject (Yarbus, 1967). One way to address the orienting task issue would be to give subjects a task that did not force the creating of a coherent memory representation for the scene, and look for similar scene context effects on fixation time across tasks.

11.5.1 Perception of Art

A particularly interesting subset of scene perception studies is the examination of gaze over a specific set of contextual images, namely art. As observed by Yarbus (1967), a viewer's intent influences eye movements and fixations over a scene. Eye movements over art have further refined the scope of these studies, examining differences in how trained viewers search for meaning and aesthetic qualities in fine art pieces.

The first systematic exploration of fixation positions in scenes was reported by Buswell (1935) (as cited by Henderson and Hollingworth (1998)), who asked 200 participants to look at 55 pictures of different types of artwork under a variety of viewing instructions. An important result was that fixation positions were found to be highly regular and related to the information in the pictures (Henderson & Hollingworth, 1998). For example, viewers tended to concentrate their fixations on people rather than on background regions when examining *Sunday on the Island of La Grande-Jatte* by Georges Seurat. These data thus provided some of the earliest evidence that eye movement patterns during complex scene perception are related to the information in the scene, and by extension, to perceptual and cognitive processing of the scene. Buswell concluded that:

> Eye movements are unconscious adjustments to the demands of attention during a visual experience. The underlying assumption in this study is that in a visual experience the center of fixation of the eyes is the center of attention at a given time.

Another example of eye movement over art studies, by Molnar (1981) (as reported by Solso (1999)), shows small differences in scanpaths between groups of subjects viewing artwork for its semantic meaning or for its aesthetic appeal. Remarkably, however, both sets of scanpaths are very similar in terms of fixated image features. In Rembrandt's *Anatomy Lesson*, fixations made by two groups of fine art students, given different sets of questions pertaining to the painting, appear to coincide on important elements even though the order of fixations may differ. In this case, the important elements are those of the faces of the anatomy professor and his students in the painting.

Further analyses of eye movements collected over art pieces reported by Molnar (as cited in Solso (1999)) seem to suggest a general rule: more complex pictures produce shorter eye fixations than less complex forms. For example, in comparison of classical works of the high Renaissance to those of the mannerist and baroque periods, classical art produces eye movements that are large and slow, reflecting the expansive nature of that style, while baroque paintings involve small and quick eye movements, reflecting the dense, animated character of that form. Baroque paintings were judged by art experts as more complex than classical paintings. It may be that complex art, which is densely packed with details, demands that attention be given to a large number of visual elements. This demand can be satisfied by allocating shorter fixation times to each feature. Simpler works may contain far fewer features vying for attention, and so may allow more time allocation per feature, resulting in longer fixations per feature.

A recent large-scale eye tracking study of art was conducted by Wooding (2002). An automated eye tracker was left running in a room of the National Gallery, London, as part of the millennium exhibit: "Telling Time". In three months, eye movements were successfully collected from 5,638 subjects while they viewed digitized images of paintings from the National Gallery collection. Since a composite representation of eye movements from so many subjects posed a problem, Wooding devised a *fixation map* method of analysis, which might be descriptively termed a landscape or terrain map of fixations, and is in fact similar to the *landscape map* developed independently by Velichkovsky, Pomplun, and Rieser (1996). The value at any point on the map indicates the height or amount of a particular property at that point (e.g., the number of fixations). An example of a fixation map and subsequent visualization of composite fixated regions are shown in Figures 11.2 and 11.3.

Future eye tracking studies over art, be they comparative or of a mass scale, will undoubtedly lead to further insights of human perception of this particularly pleasing set of images. In a similar goal, but being approached from a

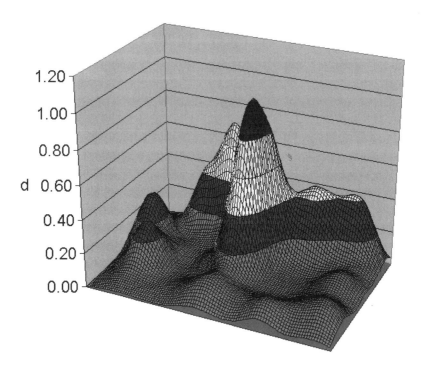

Fig. 11.2. Fixation map from 131 subjects viewing Paolo Veronese painting *Christ Addressing a Kneeling Woman*. From Wooding (2002) © 2002 ACM, Inc. Reprinted by permission.

different starting point, eye movement and general perceptual principles are currently being applied to the *generation* or creation of art by computers. A recent collaborative gathering of computer graphics and other scientists, such as representatives from perceptual and cognitive fields, met in Utah (McNamara & O'Sullivan, 2001). There, perceptual principles and eye tracking methodologies were discussed as a possible aid to the creation of more aesthetically pleasing or more interactive works of computer graphics and art.

11.5.2 Perception of Film

Another interesting form of artistic media is film. In a sense, film is a dynamic form of art. An interesting example of an eye tracking film study is given by d'Ydewalle, Desmet, and Van Rensbergen (1998). d'Ydewalle et al distinguish three levels of film editing errors in sequencing successive shots. First-order editing errors refer either to small displacements of the camera position or to small changes of the image size, disturbing the perception of apparent movement and leading to the impression of jumping. Second-order editing errors follow from a reversal of the camera position, leading to a change of the left-

(a) Original image.

(b) Visualization of important regions.

Fig. 11.3. Sample fixations from 131 subjects viewing Paolo Veronese *Christ Addressing a Kneeling Woman* © National Gallery, London, with annotations © IBS, University of Derby, UK. From Wooding (2002) © 2002 ACM, Inc. Reprinted by permission.

right position of the main actors (or objects) and a complete change of the background. With third-order editing errors, the linear sequence of actions in the narrative story is not obeyed. d'Ydewalle et al's experiment shows that there is an increase of eye movements from 200 to 400 ms following both second- and third-order editing errors. Such an increase is not obtained after a first-order editing error, suggesting that the increase of eye movements after second- and third-order editing errors is due to postperceptual, cognitive effects.

11.6 Visual Search

The question of how humans perceive the visual scene through the movements of the eyes, in the context of more natural or free tasks such as picture viewing (which is significantly different from the task of reading), can generally be modeled by the process known as visual search. Visual search, in general, refers to the process of visually scanning a scene and forming a conceptual "image" or notion of the scene as assembled by the brain. In comparison to reading, there have not been nearly as many studies dealing with visual search (Rayner, 1998).

When eye movements are recorded during extended search, fixations tend to be longer than in reading. However, there is considerable variability in fixation time and saccade length as a function of the particular search task (Rayner, 1998). Specifically, visual search tasks vary widely, and tasks in which eye movements have been monitored consist of at least the following: search (a) through text or textlike material, (b) with pictorial stimuli, (c) with complex arrays such as X-rays, and (d) with randomly arranged arrays of alphanumeric characters or objects. Because the nature of the search task influences eye movement behavior, any statement about visual search and eye movements needs to be qualified by the characteristics of the search task.

Henderson and Hollingworth (1998) list factors that may vary from study to study: image size (usually measured in visual angle), viewing task (e.g., later recognition memory task, image preference, counting non-objects, visual search, "free viewing"–it is well known that viewers place their fixations in a scene differently depending on viewing task, viewing time per scene (ranging from very short, tachistoscopic to longer durations, e.g., on the order of 50 ms–10 s or even 30 minutes (Yarbus, 1967)), image content (e.g., artwork, "natural scenes", human faces), and image type (e.g., highly artificial and regular such as sine wave gratings, color, monochrome, grayscale, or full color computer graphics imagery). These factors could each produce main effects and could also interact with each other in complex ways to influence dependent measures of eye movement behavior such as saccadic amplitudes, fixation positions, and fixation durations.

As demonstrated by Yarbus (1967) and then by Noton and Stark (1971a, 1971b), eye tracked scanpaths strongly suggest the existence of a serial component to the picture viewing process. However, serial scanpaths do not adequately explain the brain's uncanny ability to integrate holistic representations of the visual scene from piecemeal (foveal) observations. That is, certain perceptual phenomena are left unaccounted for by scanpaths, including perception of illusory images such as the Kanizsa (1976) figures or the Necker (1832) cube. While scanpaths cast doubt on a purely Gestalt view of visual perception, it would seem that some sort of holistic mechanism is at work which is not revealed by eye movements alone. Models of visual search attempt to answer this dilemma by proposing a parallel component, which works in concert with the serial counterpart exhibited by eye movements.

In visual search work, the consensus view is that a parallel, pre-attentive stage acknowledges the presence of four basic features: color, size, orientation, and presence and/or direction of motion (Doll, 1993; Wolfe, 1993). Todd and Kramer (1993) suggest that attention (presumably in the periphery) is captured by sudden onset stimuli, uniquely colored stimuli (to a lesser degree than sudden onset), and bright and unique stimuli. There is doubt in the literature whether human visual search can be described as an integration of independently processed features (Van Orden & DiVita, 1993).

Visual search, even over natural stimuli, is at least partially deterministic, rather than completely random (Doll, Whorter, & Schmieder, 1993). Determinism stems from either or both of two kinds: the observer's strategy determines the search pattern (as in reading), and/or the direction of the next saccade is based on information gained through peripheral vision about surrounding stimuli. Features that are likely to be fixated include edges, corners, spatially high frequency components, but not plain surfaces.

A theory that consolidated the serial and parallel counterparts of visual attention was put forth by Treisman (1986) (see also Treisman and Gelade (1980)). Treisman and her colleagues reported on the "pop-out" effect, that is, a visual target pops out from a field of similar but distinct distractors, seemingly drawing attention to itself (see Figure 11.4). This observation led to the proposition that the visual system can search for certain stimulus properties in parallel over the whole visual field–a pre-attentive process. Other target items that do not pop out must be searched for by the observer in some (usually serial) order–focused attentional processing. The difference between pre-attentive and post-attentive modes of processing can be demonstrated experimentally via paradigms designed to elicit visual search. In a typical visual search task, the observer is shown an array of items and asked to search for a specific target among the field of distractors. Reaction time is usually measured when reporting the presence or absence of a target. If reaction time does not vary with the number of items, the whole array is thought to be searched in parallel, or pre-attentively. If reaction time increases with the number of items, the array is most likely being scanned item by item in a serial process which is depicted by scanpaths. Treisman proposed the Feature Integration Theory (FIT) unifying these processes.

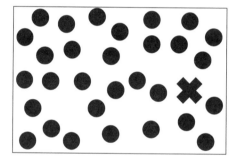

Fig. 11.4. Example of "pop-out" effect.

The main tenet of FIT is that certain features pop out because they are explicitly represented by low-level vision. This is consistent with known retinotopic maps of cortical cells tuned to basic stimulus properties such as color, orientation, and size. Conjunctions of features, however, require serial search because they do not form part of this initial representation. Thus, according to this model, visual processing can be described by three broad stages: (1) feature extraction, (2) feature binding, and (3) object representation.

Feature Integration Theory is a theory of how elementary visual features are attentionally bound together to construct unitary perceptual objects. The first (parallel) stage of processing involves multiple, retinotopic master maps of locations of elementary features, showing *where* all feature boundaries are located, but not *what* those features are. Maps correspond to feature attributes such as color, orientation, etc. According to FIT, perceptual objects are constructed by attention to a specific location, binding together the simultaneous activity at the location in all feature maps. Thus, attention can be thought of as a "glue" which integrates separated feature attributes at a particular location so that the conjunction is perceived as a unified whole.

FIT is a useful theory since it adequately explains both serial and parallel components of visual search. Parallel search is supported by FIT since targets defined by elementary properties are available in feature maps. Serial search is supported since search for conjunctions requires focused attention to bind together features in separate maps. However, while FIT can explain search performance fairly well, particularly over simple stimuli (e.g., search for Q in a field of Os), it is not clear whether FIT generalizes to more complex stimuli such as natural imagery (e.g., aerial photographs).

Feature Integration Theory is currently held to be a useful heuristic starting point for theoretical treatments of visual search although it is widely recognized to be over-simplified (Findlay & Gilchrist, 1998). Searches for more sophisticated accounts have taken various forms. Wolfe (1994) has pointed out the weakness of the assumption that search must be either serial or parallel and developed a model of the way in which interactions between serial and parallel process could occur. A particular problem of FIT is that it suggests that while pre-attentive feature processes can perform feature searches in parallel, attention from item to item is required for all other (serial) searches. Conjunction searches, however, appear to be too efficient to be explained as purely serial attentive searches. Wolfe's Guided Search (GS) model accounts for this efficiency by proposing that pre-attentive feature processes could guide the deployment of attention in conjunction searches (Wolfe, 1993, 1994). No pre-attentive process can identify a particular conjunction, however, two different pre-attentive processes (e.g., a color and an orientation process) can cooperate to mark conjunctive targets. If the output of these two pre-attentive processes is combined into an attention-guiding activation map, attention will be guided to conjunction items (e.g., black vertical bars in a field of white or black horizontal or vertical bars). Wolfe's model is discussed further in Section 11.6.1.

Relatively few studies have addressed the relationship between eye movements and the search process (Findlay & Gilchrist, 1998). Findlay and Gilchrist argue that the tradition in search research to pay little attention to eye movements and instead to use the concept of covert visual attention movements (redirecting attention without moving the eyes; an important component of FIT) is misguided. The authors demonstrate that when viewers are free to move their eyes, no additional covert attentional scanning occurs. They show that unless instructions explicitly prevent eye movements, subjects in a search task show a natural propensity to move their eyes, even in situations where it would be more efficient not to do so. The authors suggest that the reason for this preference is that in naturally occurring search situations, eye movements form the most effective way of sampling the visual field.

Findlay (1997) recorded eye movements during tasks of a simple feature search and a color shape feature conjunction search. Eye movements were recorded by having the subject wear a contact lens-type search coil positioned at the center of two large Helmholtz field coils. The induced currents in the eye coils measured eye position in space in a way which minimized head movement artifact, measuring eye position with an accuracy of 10 min arc or better following calibration. Using a single feature (color) search task, in the homogeneous distractor condition, only 0.5% of first saccades were directed at a non-target and in the heterogeneous distractor condition the percentage of misdirected saccades is under 2%. This accuracy is achieved with no cost in the time needed to program the saccade. This provides an impressive confirmation that search for a pre-specified color target can be carried out in parallel. When two targets are presented simultaneously in neighboring positions, the first saccade is directed towards some "center of gravity" position. Results suggest that the control of the initial eye movement during both simple and conjunction searches is through a spatially parallel process.

Results from a conjunction search experiment (color and shape) are particularly relevant since this is a situation where serial scanning would be expected according to the classical search theory of Treisman and Gelade (1980). If a rapid serial scanning with covert attention could occur before the saccade is initiated, it is not clear why incorrect saccades would occur as frequently as observed in the experiment (three subjects were able to locate targets with a single saccade on 60-70% of occasions in the inner ring of a two-ring concentric display, and 16-40% in the outer ring). Moreover, Findlay argues that the data place constraints on the speed of any hypothetical serial scanning process since it would be necessary for a number of locations to be scanned before the target is located, given the accuracy obtained. Alternative accounts of the visual search process have appeared which assign much more weight to the parallel processes and avoid the postulation of rapid serial scanning. The results of the conjunction search are consistent with a search model which limits parallel scanning to about 8 items but requires serial search for displays of larger number.

Bertera and Rayner (2000) had viewers search through a randomly arranged array of alphanumeric characters (i.e., letters and digits) for the presence of a target letter. They used both the moving window technique and the moving mask technique. The number of items in the array was held constant, but the size of the display varied ($13° \times 10°$, large, $6° \times 6°$, medium, and $5° \times 3.5°$, small).

In the moving window study, search performance reached asymptote when the window was $5°$. The moving mask had a deleterious effect on search time and accuracy, indicating the importance of foveal vision during search; the larger the mask, the longer the search time. For the window conditions, six different window sizes were used. The window was $1.0°$, $2.3°$, $3.7°$, $5.0°$, or $5.7°$; in addition, a control condition was run in which the entire array was presented (i.e., there was no window). Six different mask sizes were also used: $0.3°$, $1.0°$, $1.7°$, $2.3°$, $3.0°$, or no mask present. A Stanford Research Institute Dual Purkinje Eye Tracker was used to record eye movements, with 5 subjects participating in the study.

Under the window condition, search performance improved (i.e., search time decreased) as the window size increased. In general, search performance reached asymptote when the window was $5.0°$. There was no effect of either window size or array size on accuracy; the subjects successfully located the target letter 99% of the time across the different conditions. Under the mask condition, as mask size increased, search time increased. A large mask was more detrimental to search than a small window. Unlike window size, the size of the mask had a significant effect on accuracy. The accuracy values were 100%, 99%, 94%, 73%, 58%, and 39% for mask sizes 0 (no mask), $0.3°$, $1.0°$, $1.7°$, $2.3°$, and $3.0°$, respectively.

Bertera and Rayner's study shows that a useful perceptual span in visual search extends to about $5°$. The moving window paradigm for scene perception and visual search is discussed further as an instance of gaze-contingent display technology in Section 14.2.

Recently, Greene and Rayner (2001) showed that familiarity with distractors around an unfamiliar target facilitates visual search. Eye movements were recorded (right eye only) by an SMI EyeLink head-mounted tracker. Eye positions were sampled at 250 Hz by an infra-red video-based system that also compensated for head movements. Gaze positions were accurate within .5°. A saccade was recorded when eye velocity exceeded 35 deg/s or eye acceleration exceeded 9500 deg/s². Results indicated comparably long, but fewer fixations when distractors are familiar, discounting the theory that unfamiliar distractors need longer processing.

Results from search studies begin to call into question the role of covert attention in the search process (Findlay & Gilchrist, 1998). Findlay and Gilchrist advocate that a reappraisal of the role of covert attention in vision is in order since visual search does not provide a rationale for the existence of the covert attentional mechanism. Search situations in which the use of covert attention is advantageous are artificial and unusual. Most search tasks will be served better with overt eye scanning, guiding the eye as well as possible from the information which is being processed in parallel over the central regions of the visual field. The authors felt able to reject with some confidence the possibility that a fast covert scan of attention takes place during fixations in visual search. Covert attention might play no role, with a purely parallel process leading to saccade destination selection. Under this assumption, the saccade would be directed to the point of highest salience in some hypothetical 'salience map'.

11.6.1 Computational Models of Visual Search

Recent advances in the research of low-level visual processes have led to the emergence of sophisticated computational models of visual search. These models, operating on a host of still (and sometimes moving) images, model the human visual system's processes from the cornea to the striate cortex. An early example of such a model was proffered by Doll et al (1993). The model contained six modules:

1. A pattern perception module (simulates HVS from cornea to visual cortex).
2. A visual search module (takes clutter into account).
3. A detection module (calculates probability of target detection given information on fixation locations from visual search module).
4. A decision module (predicts trade-off between detections and false alarms).
5. An output/executive module (calculates cumulative probability of target acquisition and "missed" acquisitions–false alarms to clutter objects).
6. A target tracking module accounting for luminance contrast.

Doll et al's model at the time only handled luminance contrast, it did not contain modules to process chromatic contrast and motion.

Wolfe's Guided Search has undergone several revisions and "upgrades" (the current version is 3.0). Previous versions of GS successfully modeled a wide range of laboratory search tasks. The most recent revisions to the model were made in an effort to handle natural images (e.g., aerial photographs). The previous version of GS ignored two important factors: first, eye movements of human observers and the interplay of covert attentional movements with overt movements of the eyes; second, the anisotropic characteristics of the retina (visual processing is much more detailed at the fovea). The newest version of GS incorporates eye movements and eccentricity effects.

The simulation of Wolfe's Guided Search starts with a greyscale image as input (see Figure 11.5). The image is processed by an array of On- and Off-Center Units, approximating retinal ganglion cell response. These provide input to Pre-Attentive Feature Maps for brightness and orientation, which model the representation of the stimulus in V1 by a complex log transform. Next, the Attention Gate allows feature information from only one object at a time to reach the higher processes such as Object Recognition. The Attentional Gate is under the control of the Activation Map. The Activation Map is a winner-take-all network that converges on a winner about every 50ms. Feedback from the identification state to the Activation Map selects or inhibits the selected item, permitting new items to eventually win access to the identification stage.

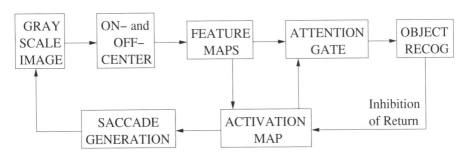

Fig. 11.5. Architecture of Guided Search 3.0. Adapted from Wolfe and Gancarz (1996) with permission © 1996 Kluwer Academic/Plenum Press.

To simulate eye movements, the Saccade Generation stage creates a saccade map. In GS, the Saccade Generation module is the analog of the superior colliculus. Activity in the saccade map is a blurred version of activity in the activation map. The GS model leads to a cooperative relationship between eye movements and attentional deployments. The simulation produces data that mimic human data on a number of tasks and GS eye movements resemble those recorded in human subjects.

Wolfe's Guided Search appears to be restricted to monochrome (grey scale) still images. A similar attentional model has recently been proposed by Osberger and Maeder (1998). Osberger and Maeder's algorithm uses an Importance Map (IM) to predict the perceptual importance of segmented image regions. The IM is built by considering both low- and high-level visual attributes of segmented regions. Low-level factors include: contrast, size, and shape. High-level factors include the region's location, and foreground/background classification. Osberger and Maeder's algorithm is designed to be flexible to allow easy accommodation of future contributing modules to the IM, including one analyzing color and motion.

An attentional model that considers color, as well as other familiar visual attributes, has been proposed by Itti et al (1998). Building on biologically-plausible architectures of the human visual system, the model is related to Treisman's Feature Integration Theory. The model architecture is shown in Figure 11.6. Starting with an input image, it is progressively low-pass filtered and subsampled to yield nine dyadic spatial scales. The multiscale image representation is then decomposed into a set of topographic feature maps. Each feature is computed via a set of linear "center-surround" operations akin to visual receptive fields. Different spatial locations then compete for saliency within each map, such that only locations which locally stand out from their surround can persist. All feature maps feed, in a purely bottom-up manner, into a master "saliency map" (SM). Feature maps are combined into three "conspicuity maps", for intensity, color, and orientation at each scale of the saliency map. The model's saliency map contains internal dynamics which generate attentional shifts. The SM feeds into a biologically-plausible "winner-take-all" (WTA) network. The Focus Of Attention (FOA) is shifted to the winning region once it has been identified by the WTA network. The WTA is then reset by a combination of local and global inhibition mechanisms which allow the selection of a future region to which the FOA is shifted in turn.

The model has been tested on a variety of artificial and natural images, and appears to be very robust, particularly to the addition of noise. Interestingly, the model was able to reproduce human performance for a number of pop-out tasks. When a target differed from an array of surrounding distractors by its unique orientation, color, intensity, or size, it was always the first attended location, irrespective of the number of distractors. Furthermore, when the tar-

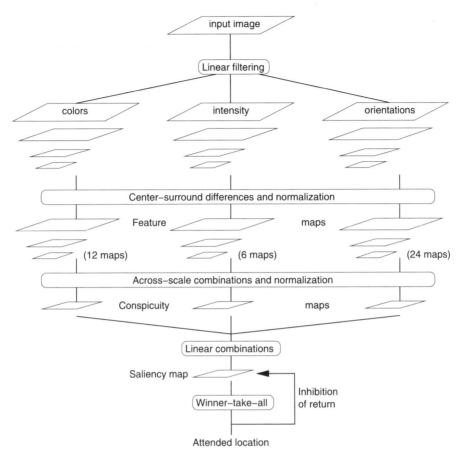

Fig. 11.6. Architecture of Itti et al's visual attention system. Adapted from Itti et al (1998) with permission © 1998 IEEE.

get differed from the distractors by a conjunction of features, the search time necessary to find the target increased linearly with the number of distractors. This performance is in general consistent with observations in humans and is consistent with Treisman's Feature Integration Theory.

Of course to test any of the above visual attention models, one very attractive methodology is to compare the sequence of Regions Of Interest (ROIs) identified by an attentional model to those actually selected by human observers. A recent study by Privitera and Stark (2000) presents just such a methodology for comparing algorithmic ROIs, or aROIs to those selected by humans, hROIs. The authors present a statistical and computational platform to perform the comparison between aROIs and hROIs. The comparison algorithm relies on two important processes: one of clustering of ROIs for comparison of loci of

ROIs and a subsequent step of assembling the temporal sequences of ROIs into ordered strings of points for comparison of sequences based on *string editing*. Clustered ROIs within an image, either viewed by human observers, or selected by an attentional algorithm, are assigned character labels. Assembled strings of ROIs are compared by string editing, which, defined by an optimization algorithm, assigns unit cost to three different character operations: *deletion*, *insertion*, and *substitution*. Characters are manipulated to transform one string to another, and character manipulation costs are tabulated to yield a sequence similarity index S_s. An example is given in Figure 11.7. A positional, or loci, similarity index S_p can be found for two strings by examining the characters of the second string to those of the first. For the strings in Figure 11.7, since all the characters of $s_2 = afbffdcdf$ are present in $s_1 = abcfeffgdc$, the two strings yield a loci similarity index of $S_p = 1$. Similarity coefficients are then sorted and stored in a table, named the Y-matrix, having as many rows and columns as the number of different sequence ROIs to be considered.

Given two strings $s_1 = abcfeffgdc$ and $s_2 = afbffdcdf$, the second can be made to equal the first by applying the following operations:

$s_1 = abcfeffgdc$		
$s_2 = afbffdcdf$	start	cost 0
$s_1 = abcfeffgdc$		
$s_2 = afeffdcdf$	substitution of first b by e	cost 1
$s_1 = abcfeffgdc$		
$s_2 = abcfeffdcdf$	insertion of bc after first a	cost 2
$s_1 = abcfeffgdc$		
$s_2 = abcfeffdc$	deletion of last df	cost 2
$s_1 = abcfeffgdc$		
$s_2 = abcfeffgdc$	insertion of g	cost 1

The total combined cost of deletions, insertions, and substitutions is 6. Relative to the original string length (9), the total cost yields a sequence similarity index between the two strings of $S_s = (1 - 6/9) = 0.34$.

Fig. 11.7. String editing example. From Privitera and Stark (2000) with permission © 2000 IEEE.

Using string editing similarity measures, Privitera and Stark evaluated six different attentional algorithms, some sharing similarities with the above models of Wolfe, Osberger and Maeder, and Itti et al. Although the algorithms tested were not the actual ones proposed by the latter group of authors, it appears that a multiresolutional strategy, such as that of Itti et al, seems to be very efficient for several classes of images. In general, although the set of algorithms picked by Privitera and Stark was only a small representative sample of many possible procedures, this set could indeed predict eye fixations.

The problem of computationally modeling human eye movements, and indeed human visual search, is far from being solved. No current model of visual search is as yet complete. Recent progress in algorithmic sophistication is encouraging. Models such as those of Wolfe, Osberger and Maeder, and Itti et al show definite promise. Certainly the capability of tracking human eye movements has yet to play a crucial role in the corroboration of any model of visual search. Privitera and Stark's approach to the comparison of human and algorithmic scanpaths is one of the first methods to appear to quantitatively measure not only the loci of ROIs but also the order of ROIs. Undoubtedly future evaluation of human and artificial scanpaths will play a critical role in the investigation of visual search.

11.7 Natural Tasks

A host of useful factual information has been derived through psychophysical testing (e.g., spatial acuity, contrast sensitivity function, etc.). These types of studies often rely on the display of basic stimuli, e.g., sine wave gratings, horizontal and vertical bars, etc. Although certainly central to the development of such theories as Feature Integration, one criticism of these artificial stimuli is their simplicity. As discussed above, studies of visual search are expanding to consider more complex stimuli such as natural scenery (Hughes, Nozawa, & Kitterle, 1996). However, viewing of pictures projected on a laboratory display still constitutes something of an artificial task. Recent advancements in wearable and virtual displays now allow collection of eye movements in more natural situations, usually involving the use of generally unconstrained eye, head, and hand movements.

Important work in this area has been reported by Land, Mennie, and Rusted (1999) and Land and Hayhoe (2001). The aim of the first study was to determine the pattern of fixations during the performance of a well-learned task in a natural setting (making tea), and to classify the types of monitoring action that the eyes perform. Results of this study indicate that even automated routine activities require a surprising level of continuous monitoring. A head-mounted eye-movement video camera was used, which provided a continuous view of the scene ahead, with a dot indicating foveal direction with an accuracy of about 1 deg. Foveal direction was always close to the object being manipulated, and very few fixations were irrelevant to the task. Roughly a third of all fixations on objects could be definitely identified with one of four monitoring functions: *locating* objects used later in the process, *directing* the hand or object in the hand to a new location, *guiding* the approach of one object to another (e.g., kettle and lid), and *checking* the state of some variable (e.g., water level). Land et al (1999) conclude that although the actions of tea-making are 'automated' and proceed with little conscious involvement, the eyes closely monitor every step of the process. This type of unconscious attention must be a common phenomenon in everyday life.

Investigating a similar natural task, Land and Hayhoe (2001) examined the relations of eye and hand movements in extended food preparation tasks. The paper compares the task of tea-making against the task of making peanut butter and jelly sandwiches. In both cases the location of foveal gaze was monitored continuously using a head-mounted eye tracker with an accuracy of about 1 deg, and the head was free to move. In the tea-making study the three subjects had to move about the room to locate the objects required for the task; in the sandwich-making task the seven subjects were seated in one place, in front of a table. The eyes usually reached the next object in the sequence before any sign of manipulative action, indicating that eye movements are planned into the motor pattern and lead each action. The eyes usually fixated the same object throughout the action upon it, although they often moved on to the next object in the sequence before completion of the preceding action. Specific roles of individual fixations were found to be similar to roles in the tea-making task (see above). Land and Hayhoe argue that, at the beginning of each action, the oculomotor system is supplied with the identity of the required objects, information about its location, and instructions about the nature of the monitoring required during the action. The eye movements during this kind of task are nearly all to task-relevant objects, and thus their control is seen primarily 'top-down', and influenced very little by the 'intrinsic salience' of objects. General conclusions provided by Land and Hayhoe are that the eyes provide information on an 'as needed' basis, but that the relevant eye movements usually precede the motor acts they mediate by a fraction of a second. Eye movements are thus in the vanguard of each action plan, and are not simply responses to circumstances. Land and Hayhoe conclude that their studies lend no support to the idea that the visual system builds up a detailed model of the surroundings and operates from that. Most information is obtained from the scene as it is needed.

A good deal of additional work on eye movement measurement during natural tasks has also been performed by a group of researchers seemingly co-located around Rochester, NY, mainly at the University of Rochester and the Rochester Institute of Technology. This group of researchers has been investigating the relationship between eye, head, and hand movements for some time (among other topics).

Ballard, Hayhoe, and Pelz (1995) investigated the use of short-term memory in the course of a natural hand-eye task. The investigation focused on the minimization of subjects' use of short-term memory by employing deictic primitives through serialization of the task with eye movements (e.g., using the eyes to "point to" objects in a scene in lieu of memorizing all of the objects' positions and other properties). The authors argue that a deictic strategy in a pick-and-place task employs a more efficient use of a frame of reference centered at the fixation point, rather than a viewer-centered reference frame which might require memorization of objects in the world relative to coordinates centered at the viewer. Furthermore, deictic strategies may lead to a computational simplification of the general problem of relating internal models to objects in the world. Sequential, problem-dependent eye movements avoid the general problem of associating many models to many parts of the image simultaneously. Thus, in a pick and place task, the steps required to perform the task of picking up a certain object from a group of objects can employ deictic references by fixating the object to be picked up next, without having to internalize a geometric reference frame for the entire set of objects.

Ballard et al tested the above deictic reference assumption to see whether humans in fact use their eye movements in a deictic fashion in the context of natural behavior. A head mounted eye tracker was used to measure eye movements over a three-dimensional physical workplace block display, divided into three areas, the *model*, *source*, and *workspace*. The task assigned to subjects was to move and assemble blocks from the source region to the workspace, arranging the blocks to match the arrangement in the model area. An example of the setup is shown in Figure 11.8. By recording eye movements during the block pick-and-place task, the authors were able to show that subjects frequently directed gaze to the model pattern before arranging blocks in the workspace area. This suggests that information is acquired incrementally during the task and is not acquired *in toto* at the beginning of the tasks. That is, subjects appeared to use short-term memory frugally, acquiring information just prior to its use, and did not appear to memorize the entire model block configuration before making a copy of the block arrangement.

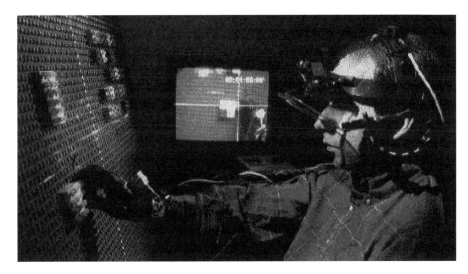

Fig. 11.8. Eye tracking in a natural pick-and-place task. Courtesy of Jeff Pelz, Visual Perception Laboratory, Carlson Center for Imaging Science, Rochester Institute of Technology <http://www.cis.rit.edu/people/faculty/pelz/research/ASL-tracker.html>. Reproduced with permission.

In a similar block-moving experiment, Smeets, Hayhoe, and Ballard (1996) were able to show that horizontal movements of gaze, head, and hand followed a coordinated pattern. A shift of gaze was generally followed by a movement of the head, which preceded the movement of the hand. This relationship is to a large extent task-dependent. In goal-directed tasks in which future points of interest are highly predictable, the authors hypothesize that while gaze and head movements may decouple, the actual position of the hand is a likely candidate for the next gaze shift.

A recent example of such intentional visual attention was demonstrated by Pelz, Canosa, and Babcock (2000). Using a wearable eye tracker, Pelz was able to show intentionally-based eye movements, which he termed "look-ahead" eye movements, during a simple hand-washing tasks. In this experiment, a subject donned a wearable computer and eye tracking rig (the computer was worn in a backpack). During a simple hand-washing task, recorded eye movements showed that gaze moved to a location (soap dispenser) prior to the movement of the hands to the same location (see Figure 11.9).

To further examine issues raised by observations of natural behavior, Hayhoe et al (2002) have recently begun using complex virtual environments which can be manipulated by the experimenter at critical points during task performance. In a virtual environment where subjects copy toy models, the authors

Fig. 11.9. Eye tracking in a natural hand-washing task. From Pelz et al (2000) © 2000 ACM, Inc. Reprinted by permission.

show that regularities in the spatial structure are used by subjects to control eye movement targeting. Other experiments in a virtual environment with haptic feedback show that even simple visual properties like size are not continuously available or processed automatically by the visual system, but are dynamically acquired and discarded according to the momentary task demands.

11.8 Eye Movements in Other Information Processing Tasks

Eye movements have been recorded and studied in a host of information processing tasks. Rayner (1998) provides a comprehensive review of eye movement work from multiple domains. The reader is referred to Rayner's article for the full account, the remaining classes of eye tracking applications not discussed above are listed here in point form. Unless otherwise noted, the information on this research comes from Rayner (1998).

Auditory Language Processing. In this paradigm, eye movements are recorded as people listen to a story or follow instructions regarding an array they are looking at. Cooper (1974) introduced this method and found that when people are simultaneously presented with spoken language and a visual field containing elements semantically related to the informative items of speech, they tend to spontaneously direct their line of sight to those elements which are most closely related to the meaning of the language currently heard. Cooper observed three main types of visual behavior: (1) a visual-aural interaction mode, in which fixation of targets was correlated with the meaning of concurrently heard language; (2) a free-scanning mode, in which the subject continually altered his direction of gaze in a manner independent of the meaning of concurrently heard language; (3) a point-fixation mode, in which the subject continued to fixate the same location independent of the meaning of concurrently heard language. It was frequently the case that subjects would vacillate between more than one of these modes during the presentation of a single story or comprehension test. Informal evidence based upon subjects' post-experimental verbal reports suggests that these three types of visual behaviors may be related to their distribution of attention between the visual and auditory modalities.

The eye movement paradigm has also been applied to auditory language processing by Allopenna, Magnuson, and Tanenhaus (1998). These authors used a paradigm in which participants followed spoken instructions to manipulate either real objects or pictures displayed on a computer screen while their eye movements were monitored using an Applied Science Laboratories (ASL) E4000 eye tracker which features a lightweight camera mounted on a headband. Allopenna et al found that eye movements to objects in the workspace are closely time-locked to referring expressions in the unfolding speech stream, providing a sensitive and nondisruptive measure of spoken language comprehension during continuous speech.

Allopenna et al (1998) addressed two important methodological issues with the eye tracking paradigm. First, they showed that the use of a restricted set of lexical possibilities does not appear to artificially inflate similarity effects. In particular, no evidence for rhyme effects was found with successive gating, which is a task that emphasizes work-initial information. Second, the authors provided clear evidence in support of a simple linking hypothesis between activation levels and the probability of fixating on a target. The predicted probability that an object would be fixated over time closely corresponded to the behavioral data. The availability of a mapping between hypothesized activation levels and fixation probabilities than can be used to generate quantitative predictions means that eye movement data can be used to test detailed predictions of explicit models.

Allopenna et al argue that the sensitivity of the response measure coupled with a clear linking hypothesis between lexical activation and eye movements indicates that this methodology will be invaluable in exploring questions about the microstructure of lexical access during spoken word recognition. Eye movement methodology should be especially well suited to addressing questions about how fine grained acoustic information affects word recognition. Allopenna et al believe that a particularly exciting aspect of the methodology is that it can be naturally extended to issues of segmentation and lexical access in continuous speech under relatively natural conditions.

Mathematics, Numerical Reading, and Problem Solving. In this area of investigation, eye movements are recorded as participants solve math and physics problems, as well as analogies. Not surprisingly, more complicated aspects of the problems typically lead to more and longer fixations.

Eye Movements and Dual Tasks. This methodology involves examination of eye movements when viewers are engaged in a dual-ask situation, for example, a speeded manual choice response to a tone is made in close proximity to an eye movement. Although there is some slowing of the eye movement, the dual-task situation does not yield the dual-task interference effect typically found.

Face Perception. When examining faces, people tend to fixate on the eyes, nose, mouth, and ears. Fixations tend to be longer when comparisons have to be made between two faces rather than when a single face is examined.

Illusions and Imagery. These studies deal with illusions, such as the Necker cube or ambiguous figures.

Brain Damage. Brain damage studies have examined eye movements of patients with scotomas and visual neglect as they engage in reading, visual search, and scene perception.

Dynamic Situations. Eye movements have been examined in a host of dynamic situations such as driving, basketball foul shooting, golf putting, table tennis, baseball, gymnastics, walking in uneven terrain, mental rotation, and interacting with computers. Some of these applications are covered in the next chapter. Studies in which eye-hand coordination is important, such as playing video games, have revealed orderly sequences in which people coordinate looking and action.

11.9 Summary and Further Reading

This chapter presented eye tracking-related work mainly related to neuropsychology. Neuroscientific investigation linking functional brain mapping and eye tracking may seem a touch futuristic, however, functional brain imaging devices combined with eye trackers are already available. Currently such devices are undoubtedly expensive, however, it is certain that in due time these devices will be used to investigate visual attentional phenomena from entirely new perspectives.

The chapter focused mainly on traditionally psychological investigations of vision featuring eye tracking: vision during reading, visual search, perception of art, and vision in natural and virtual environments. As can be seen, there is a good deal of potential for collaboration between computer scientists, eye tracking researchers, and scientists investigating visual perception. Providing the capability of recording eye movements over new and complex stimuli will undoubtedly extend our knowledge of vision and visual perception.

Good sources of information for keeping track of this type of work are scientific journals and conferences dealing with vision, psychology, and eye tracking. Examples include Vision Research, Behavior Research Methods, Instruments, and Computers (BRMIC), and the proceedings of the European Conference on Eye Movements (ECEM), and the US-based Eye Tracking Research & Applications (ETRA).

12. Industrial Engineering and Human Factors

Eye tracking offers a unique measure of human attentional behavior. This is particularly important in evaluating present and future environments in which humans do and will work. The examination of human interaction and behavior within their environments, particularly ones in which humans often perform functions critical to safety, is one topic studied by human factors and industrial engineers. Traditional measurement methods of human performance often include measures of reaction time and accuracy, e.g., how fast a person completes a task and how well this task is performed. These are generally measures associated with performance. To study the steps taken to perform the tasks requires analysis of the individual procedures performed. For this analysis, process measures are often needed. Eye movements are particularly interesting in this latter context since they present measures which can provide insights into the visual, cognitive, and attentional aspects of human performance.

Here, three broad experimental domains are presented in which eye tracking can play an important analytical role: aviation, driving, and visual inspection. With the development of sophisticated simulators for this task, and incorporation of eye trackers into these simulators, analysis of eye movements provides a powerful additional mechanism for measuring human factors.

12.1 Aviation

Since their early use in flight simulators (e.g., see Kocian (1987), Longridge, Thomas, Fernie, Williams, and Wetzel (1989)), eye trackers have continued their utility in aviation experiments, ranging from testing procedural training to the evaluation of increasingly sophisticated deployment of new (graphical) displays. With improved simulator and eye tracking technology, the use of gaze recording techniques will undoubtedly continue.

An example of a recent combined use of relatively new eye tracking technology in a sophisticated flight simulator was reported by Anders (2001). Eye and head movements of professional pilots were recorded under realistic flight conditions in an investigation of human-machine interaction behavior relevant to information selection and management as well as situation and mode awareness in a modern glass cockpit. The simulator used was the Airbus 330 full flight simulator certified for airline training. Microphones and video cameras inside the simulator record environmental sounds, voice communication, general cockpit settings and the actions of the operators. Several flight trials were conducted, with each flight starting from level 210 (21,000 feet altitude) in managed descent mode including an ILS (Instrument Landing System) approach and landing. The captain's eye and head movements were recorded during trials for later Point Of Regard (POR) analysis within a previously defined 3D model of the environment. Areas of interest were chosen to augment eye movement analysis. Major displays and panels were selected, including Primary Flight Display (PFD), Navigation Display (ND), Engine and Warning Display (EWD), System Display (SD), Main Panel (MPn), Glare Shield (GSc), and Overhead + Central Pedestal (OCA). The Flight Control Unit (FCU) is part of the GSc whereas the two Multipurpose Control and Display Units (MCDU) in front were separately defined as CDU. MAP described the chart and other paper information, with OUT representing the window area. The pre-defined cockpit regions of interest are shown in Figure 12.1.

Analysis of eye movements illustrates the importance of the PFD as the primary source of information during flight (the PFD is the familiar combined artificial horizon and altimeter display gauges for combined altitude and attitude awareness). Within the PFD, most of the fixation time is attributed to flight parameters presented on the speed band (SPD), artificial horizon (AH), and altitude band (ALT). Cumulative fixation time on the heading band (HDG) is fairly low since the ND gives a better indication of horizontal situation. Average attention allocation to areas of interest is very similar for all tested approach flights. Between-subject differences (across pilots) are reportedly not outstanding. Attention shifts from the PFD to OUT shortly before landing. Attentional shifts were also recorded to the GSc in response to simulated Air Traffic Control (ATC) instructions to change flight parameters. Attention is high to the MAP and CDU areas during phases of flight in which the pilot is briefing the approach or when the flight plan changes (e.g., after a "Go direct..." or runway change instructions from ATC). As a proof of concept, this study shows the potential of eye movements for judgment of pilots' performance and future training of novice pilots.

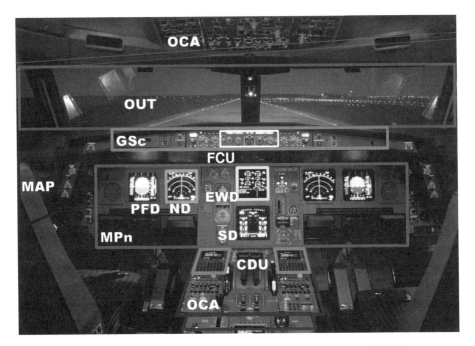

Fig. 12.1. A330 cockpit with pre-defined areas of interest. Courtesy of Geerd Anders.

Besides general scanpath patterns in the cockpit, eye movements have also been used to evaluate the usability of specific instruments such as newly developed electronic maps. Electronic maps such as 3D graphical navigational displays are becoming increasingly viable alternatives to traditional maps in the cockpit (Ottati, Hickox, & Richter, 1999). In their study, Ottati et al compared eye movement patterns on different terrain features between experienced and novice pilots during a Visual Flight Rules (VFR) simulation. Based on previous studies of pilots' eye movements during Instrument Flight Rules (IFR) navigation, the authors expected experienced aviators to spend less time finding and fixating on their navigational landmarks, while novices were expected to have greater difficulty finding landmarks and extracting useful data from them, causing greater dwell times. As expected, a greater tendency in novice pilots was found to "fly out of the window" (i.e., devoting visual attention outside the cockpit) than in experienced pilots. Independent of expertise, pilots spent more time out of the window than over instruments.

The primary difference between experienced and novice pilots' eye movements during VFR flight reported by Ottati et al was the amount of significant information-gathering fixations exhibited by each group. Experienced pilots employed significantly more fixations, yet each fixation lasted for relatively the same duration. This may indicate that novices were ineffectively scanning their forward field of view (outside the window) to gain sufficient navigational information. Experienced pilots' performance suggests that their fixations out of the window were more deliberate and informative. The authors suggest that if novices were trained in experienced pilot scan strategies, the evident performance gap between novices and experts could possibly be reduced.

In a study of electronic maps for taxiing, Graeber and Andre (1999) also suggest that training is necessary to assure proper usage of, and optimal visual attention interaction with Electronic Moving Maps (EMMs). The EMM studied by the authors is a display developed by NASA to increase navigation situation awareness, decrease navigation errors, increase forward taxi speeds, decrease planning time, decrease navigation mental workload, and improve navigation communications of pilots in low-visibility conditions. The main objective of Graeber and Andre's study was to understand how pilots visually interact with the EMM. This objective was partially motivated by concerns of EMMs disproportionately drawing the pilot's eyes into the cockpit during taxi, since taxiing is a primarily "eyes-out" task. Pilots navigate the airport and terminal areas using visual cues and signage from the outside environment, and through communication with ground control. Their task is to accurately navigate the cleared route, monitor for potential incursions, and maintain a safe distance from other aircraft, ground vehicles, and obstacles—a task best served by keeping their heads-up and eyes-out the cockpit windshield.

To examine pilots' visual interaction with the EMM and outside environment, Graeber and Andre examined pilots' eye movements under three different visibility conditions: daytime high visibility (Day Visual Meteorological Conditions, VMC), daytime low visibility (Day 700 Runway Visual Range, or RVR), and nighttime moderate visibility (Night 1,400 RVR). Results indicate that visibility significantly affected the amount of time pilots dwelled on the EMM, with dwell time significantly higher in Day VMC than in Day 700 RVR (i.e., under high visibility conditions). That is, as visibility degrades, pilots spend more time eyes-out and less time dwelling on the EMM with no loss in taxi performance. A potential explanation for this surprising result is that pilots need to be eyes-out to maintain lateral and directional loop closure, scan for hazards, and maintain information gathering Out The Window (OTW). This explanation is supported by the increased percent time on route found in the performance data as the visibility decreased. As intended in the design of the EMM, it appears that pilots used the EMM as a secondary navigation aid across all visibility conditions. That is, when visibility degrades, the EMM appears to benefit performance the most, but without incurring additional visual attention requirements.

12.2 Driving

Another type of simulator in which eye trackers are increasingly being incorporated is the driving simulator. It is widely accepted that deficiencies in visual attention are responsible for a large proportion of road traffic accidents (Chapman & Underwood, 1998). An understanding of the visual search strategies of drivers is thus extremely important, and much research has been conducted in this area. Several important driving studies incorporating an eye tracker are described in Underwood (1998).

Eye movement recording and analysis provide important techniques for understanding the nature of the driving task and are important for developing driver training strategies and accident countermeasures (Chapman & Underwood, 1998). Chapman and Underwood indicate that at their simplest, drivers' fixation patterns on straight roads can be described as concentrating on a point near to the focus of expansion (the point in the visual field in front of the driver where objects appear stationary) with occasional excursions to items of road furniture and road edge markers. This reliance on the focus of expansion in the scene is assumed to be because it provides precise directional information to the driver and is the location near to which future traffic hazards are likely to be first visible. Increasing the complexity of the visual scene (by adding vehicles, road furniture, or irrelevant signing) increases the number of eye movements made and decreases the mean fixation durations on individual objects. This seems to be a natural response to having more objects available in the visual field to look at; it is not clear whether decreases in fixation durations mean that objects are processed less completely or that redundant fixation time is simply reduced.

Chapman and Underwood's study reports on recorded eye movements of relatively large groups of both novice and experienced drivers while watching videos of dangerous situations. The largest overall difference between the groups was that novices had generally longer fixation durations than experienced drivers in this task. Chapman and Underwood argue that this reflects the additional time required by novices to process information in the visual scene. Important differences were also found between the types of situation used in the study. Rural situations, even those chosen to be particularly dangerous, generally evoked fewer responses from subjects and longer fixation durations. In urban films both groups of subjects reported many more dangerous events and had shorter mean fixation durations. It is clear that dangerous events generally evoke long fixation durations. This result demonstrates the dangers in averaging eye movement data over periods of time in dynamic scenes. It is therefore argued that to fully understand the subtlety of such data, and to draw realistic conclusions about the cognitive process underlying observable behaviors, it is necessary to develop a detailed understanding of the moment by moment "syntax" of driving situations.

Dishart and Land (1998) discuss previous work showing that experienced drivers obtain visual information from two sections of their view of the road ahead, in order to maintain a correct position in lane whilst steering their vehicle around a curve. The more distant of these two sections is used to predict the road's future curvature. This section is optimally 0.75–1.00 s ahead of the driver and contains the tangent point. This section of road is used by a feedforward (anticipatory) mechanism which allows the driver to match the curvature of the road ahead. The other, nearer, section is about 0.5 s ahead of the driver and is used by a feedback (reactive) mechanism to "fine tune" the driver's position in lane. As either lane edge approaches the vehicle, the driver steers away from it, correcting his/her road position. Experiments using video based eye-head tracking equipment have shown that the feedback mechanism is present in most people regardless of their driving experience (although its accuracy is higher in those with experience), but that the feedforward mechanism is learned through experience of steering tasks (that can include riding a bicycle, computer driving games, etc.).

Dishart and Land (1998) discuss eye-head tracking experiments on learner drivers during their tuition which indicate that use of the section of the road containing the tangent point increases with experience, then decreases as drivers learn to optimize their visual search patterns. This affords experienced drivers to spend more of their visual resources on other visual tasks both related and unrelated to driving.

A fundamental problem in visual search in driving research is defining and controlling demand on the driver as an independent variable (Crundall, Underwood, & Chapman, 1998). Possible confounding factors are increases in visual demand, such as an increase in visual clutter or complexity, and increases in cognitive demand, such as an increase in the processing demands of a particular stimulus perhaps due to an increase in its relevance to a current context. Despite the lack of consistency in the manipulations of visual demand, it seems fairly well documented that general increases in task demands and visual complexity tend to reduce mean fixation duration and increase the sampling rate. Crundall et al discuss the effects of cognitive and visual demand upon visual search strategies during driving, and whether they can be used to differentiate between novices who have recently passed their test and more experienced drivers. The authors review previous studies with the aim of identifying a process which may account for the effects of changes in demand according to driver experience.

In a driving study examining the effects of clutter, luminance, and aging, but not quite in a driving simulation, Ho, Scialfa, Caird, and Graw (2001) recorded subjects' visual search for traffic signs embedded in digitized images of driving scenes. Example driving stimulus images are shown in Figure 12.2. In a visual search task, a traffic sign was first presented to subjects, followed by a scene in which subjects were instructed to search for the sign. Analysis of five dependent measures are reported: (a) errors, (b) reaction time, (c) fixation number, (d) average fixation duration, and (e) last fixation duration. For each measure the effects of age, clutter, luminance, and target presence were evaluated. On average, older adults were less accurate than younger ones. Accuracy for daytime scenes was independent of target presence, but errors for nighttime scenes were more common on target-present trials than on target-absent trials. In daytime scenes errors were generally more common in high clutter than in low clutter, while in nighttime scenes no consistent clutter effect was detected. Reaction time data indicated that younger adults generally responded more quickly than older adults. Reaction times were also dependent on the amount of clutter and the presence of the target. Fixation number data, a measure that is strongly correlated with reaction time, indicated that older adults made more fixations than did younger adults, and more fixations were needed for high-clutter and target-absent scenes than for low-clutter and target-present scenes. Nighttime scenes with high-clutter also required more fixations than did daytime scenes. Examining average fixation duration, it was found that younger adults had shorter fixation durations than did older adults, and low-clutter scenes resulted in shorter fixations than did high-clutter scenes. Relative to daytime scenes, high-clutter nighttime scenes containing a target resulted in longer average fixation durations. The last fixation duration reflects the comparison of the last fixated object with the target representation and the terminal decision regarding target presence. Age differences in this last stage of processing might accrue because of difficulties with the comparison process itself or because the elderly are more cautious in making overt responses. This data showed that age differences were greater on target-present trials than on target-absent trials. In summary:

- High-clutter scenes required longer latencies and more fixations to acquire the sign, were associated with more errors, and had longer fixation durations.
- Luminance effects were negligible, although high-clutter nighttime scenes did impair search efficiency on reaction time and fixation number measures. Additionally, search was impaired on target-absent trials for both age groups.
- Older adults were found to be less accurate and slower than younger adults and executed more eye movements to acquire signs. The age effects on reaction time and fixation number were more pronounced on target-absent trials. Older adults used the visual cues that discriminated targets and distractors to quickly isolate the target on target-present trials. However, they took longer to correctly decide that a target sign was not present on target-absent trials, and they also took longer to make a terminal decision regarding the sign's match to the trial target on target-present trials. Interestingly, older adults were not adversely affected by increased clutter than were the young.

Considering the rapidly aging driving population, the authors recommend that roadway engineers should consider reducing the number of competing signs (e.g., advertisements), avoiding redundant signs, and making traffic signs that are central to the safety of the driver more conspicuous.

Experiments where gaze is monitored in a simulated driving environment demonstrate that visibility of task relevant information depends critically on active search initiated by the observer according to an internally generated schedule, which depends on learnt regularities in the environment (Hayhoe et al, 2002). The driving simulator used by Hayhoe et al. consists of a steering station mounted on top of a 6 Degree Of Freedom (6DOF) hydraulic platform. Steering, accelerator, and brake are instrumented with low-noise potentiometers that modulate signals read by a special-purpose 12-bit A/D board on an SGI VME bus. Subjects drive in "PerformerTown", an extensible environment designed by Silicon Graphics. Cars and trucks have been added to the environment that move along predefined or interactively controlled paths. The visual environment is easily modified to simulate lighting at any time of the day or night, along with the addition of fog. Observers view the display in a Head Mounted Display (HMD) and eye position is monitored using a video-based corneal reflection eye tracker. This simulator has been used to examine "change blindness", a phenomenon where observers are insensitive to many changes made in scenes either during a saccadic eye movement or some other masking stimulus. In one experiment, while subjects drove along a pre-specified path following a lead car, in a particular block a No Parking sign changed into a

(a) Daytime.

(b) Simulated nighttime.

Fig. 12.2. High clutter driving stimulus images. Reprinted with permission from *Human Factors*, Vol.43, No.3, 2001. Copyright 2001 by the Human Factors and Ergonomics Society. All rights reserved.

Stop sign for about 1 second as the observer approached it. The retinal transient caused by the change was masked by 100ms blank screens before and after the change. Results indicate that the likelihood of fixating the sign at any point while it is there is heavily modulated by the task and by the location of the sign. Given a "follow" instruction, observers rarely fixate the sign. When driving normally they invariably respond when it is at an intersection, but miss it 2/3 of the time when it is in the middle of the block. This suggests observers must actively initiate a search procedure in order to see particular stimuli in the scene. Markedly different fixation patterns were displayed in the two instructions. In the "drive normally" conditions, subjects spent much more time fixating in the general region of the intersection. This strongly suggests fixation patterns and attentional control in normal vision is learnt. Furthermore, the visibility of traffic signs depends on active search according to an internally generated schedule, and this schedule depends both on the observer's goals and on learnt probabilities about the environment. The experiment suggests that a fuller understanding of the mechanisms of attention, and how attention is distributed in a scene, needs to be situated in the observer's natural behavior.

Research on the effects of mental activity during driving suggests the convenience of raising drivers' awareness about the possible consequences of driving while their attention is focused on their own thoughts, unrelated to driving (Recarte & Nunes, 2000). Recarte and Nunes studied the consequences of performing verbal and spatial-imagery tasks on visual search when driving. On each of 4 routes (2 highways and 2 roads), participants performed 2 verbal tasks and 2 spatial-imagery tasks while their eye movements were recorded. The same results were repeated on all routes. Pupillary dilation indicated similar effort for each task. Visual functional-field size decreased horizontally and vertically, particularly for spatial-imagery tasks. Compared with ordinary driving, fixations were longer during the spatial-imagery task. With regard to driving performance, glance frequency at mirrors and speedometer decreased during the spatial-imagery tasks. Performing mental tasks while driving caused an increased attentional workload on ordinary thought, as shown by pupillary dilation. Mental tasks imposed by the experimenter while the participant was driving, and, therefore, of a more mandatory nature than ordinary thoughts, produced: (a) marked changes in the visual inspection patterns; (b) qualitatively different changes, depending on the type of processing resources required by the mental tasks; (c) the same effects in the four different driving scenarios; and (d) changes in practical drivings behaviors, such as inspection reduction of mirrors and speedometer. Spatial-imagery tasks produced more marked effects in almost all the analyzed variables than verbal tasks. First, an increment in mean fixation durations was found, due to some long fixations and possibly associated with mental image inspection as part of the mental activity of image search or rotating. It is suggested that these eye freezing responses produce impairment of environment perception. Second, a marked reduction of the visual inspection window was found, both horizontally and vertically, possibly associated with narrowing of the attentional focus size. Smaller saccadic size and marked reduction of glance frequency at mirrors and speedometer were also noted. With regard to the implications for driving, Recarte and Nunes suggest that the spatial reduction of the visual inspection window, including the reduction of the inspection of mirrors, could be interpreted as a predictor of decreased probability of detecting traffic events, particularly when performing mental spatial-imagery tasks.

It has been experimentally demonstrated that the pattern of eye fixations reflects, to some degree, the cognitive state of the observer (Liu, 1998). Liu suggests that the analysis of drivers' eye movement may provide useful information for an intelligent vehicle system that can recognize or predict the driver's intention to perform a given action. Such a system could improve the interaction between the driver and future vehicle systems and possibly reduce accident risk. Liu reviews numerous studies that have tried to establish the level at which the relationship between eye movements and higher cognitive processes can be modeled. The least controversial conclusions simply state the the pattern of eye movements generally reflect the observer's thought processes, indicating to some degree the goals of the observer and perhaps even the main areas of interest. The strongest conclusions assert that the eye movements are directly observable indicators of underlying cognitive processes, revealing the nature of the acquired information as well as the computation processes. To implement a "smart car", one which would be able to predict or recognize the driver's intentions and then take the appropriate course of action based on that prediction, Liu suggests that it may be possible to determine the current driver state (e.g., centering of the car in the current lane, checking whether the adjacent lane is clear, steering to initiate a change in heading, centering the car in the new lane, etc.) from observed eye movement patterns via hidden Markov dynamic models (HMDMs). The primary consideration for including driver eye movements in HMDMs is how the eye movement behavior will be coded (e.g., by gaze locations or by gaze or saccade direction). Gaze location seems to be the natural choice given that characteristic patterns of driver eye movements have been identified using a Markovian analysis of gaze location. The results from preliminary experiments using HMDMs without utilizing eye movements are promising. The addition of eye movement analysis should enhance the system performance by enabling recognition of driver intentions rather than just recognition of maneuvers as they begin.

12.3 Visual Inspection

In a study of visual inspection of integrated circuit chips Schoonard, Gould, and Miller (1973), state that "visual inspection pervades the lives of all people today. From poultry, meat, and fish inspection, to drug inspection, to medical X-ray inspection, to production line inspection, to photo interpretation, the consequences of inspection directly affect people's lives through their effects on the quality and performance of goods and services."

Important criteria of evaluating the efficiency of human visual inspection include measurements of performance, as indicated by the inspector's speed and accuracy. These *performance measures* effectively summarize general outcomes of the process of (visual) inspection. In contrast, eye movements captured during visual inspection provide visualization of the inspector's process, and therefore provide an instance of *process measures*. Important eye movement related process measures include fixation durations, number of fixations, inter-fixation distances, distribution of fixations, fixational sequential indices (order of fixations) and direction of fixations. The utility of process measures relies on their relation to, or explanation of performance. Thus one expects to find a relation between process and performance measures.

With respect to performance, a key issue is the ability to gauge the visual search process with respect to some identifiable baseline measurement. For example, for training or evaluation purposes, it is desirable to be able to compare an inspector's visual search process to that of an expert inspector, or to some ideal model of performance. Visual search can be described by a two-stage process:

1. A time-consuming visual search for a target.
2. A decision process identifying the found item as either target or non-target.

Visual search can be modeled by two idealized processes: a perfect-memory systematic (serial) search, or a memory-less random search. A systematic search differs from a random search principally in the manner of visitation of searchable regions. That is, systematic search, due to its perfect memory, will visit each portion of the entire search region only once. In contrast, under random search, every portion of the search area has equal probability of being visited at any time during the search. Since there is no memory of past visitations, regions may be visited more than once (re-inspected).

To model random visual search more formally, consider a visual search area, A, with randomly distributed search targets (symbolized by $\times$), and a a visual lobe (instantaneous fixation size) with area a, as shown in Figure 12.3. Let the probability of detecting a target (e.g., a defect) given a fixation at a particular location be s, i.e.,

$$p(\text{detecting defect} \mid \text{fixation}) = s, \quad s \in (0,1)$$

Then let the probability that a fixation contains (or lands on) a target be $n_d * a/A$, i.e.,

$$p(\text{fixation contains a defect}) = n_d * \frac{a}{A}$$

where n_d is number of randomly distributed defects. Thus, the probability of detecting a defect is:

$$p(\text{detecting a defect}) = n_d * \frac{a}{A} * s = Q$$

on any single fixation. The probability of detecting a defect on any i^{th} fixation, in a series of fixations, is:

$$p(\text{detecting a defect on } i^{th} \text{ fixation}) = (1 - Q)^{i-1}Q$$

where $(1 - Q)$ is the probability of not detecting the fault on previous fixations. Thus the cumulative probability of detecting a defect over n fixations is:

$$Q \sum_{k=1}^{n} (1-Q)^{k-1} = [1 - (1-Q)^n] = 1 - e^{-Qn}$$

$$= 1 - e^{\left[n_d * \frac{a}{A} * s * n\right]}$$

Considering the time needed to search, t, and the time devoted per fixation, t', letting $n = t/t'$, the cumulative probability can be rewritten as:

$$Q = 1 - e^{\left[n_d * \frac{a}{A} * s * t / t'\right]}$$

$$= 1 - e^{-\lambda t}, \quad \text{with } \lambda = n_d * \frac{a}{A} * \frac{s}{t'}$$

Q is an exponential distribution which approaches probability 1 as search time increases. A stylized plot of the exponential distribution is shown in Figure 12.4 (lowest solid curve). The model essentially predicts that with more time, eventually all defects (or almost all defects) will be detected. An important attractive feature of this model is that it predicts the accuracy of detection during the first few seconds of search, meaning that the probability of detection

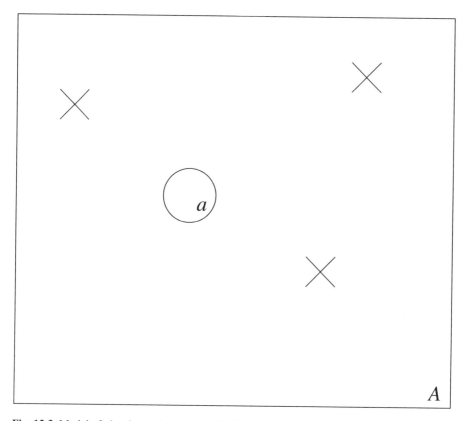

Fig. 12.3. Model of visual search area, visual lobe, and targets.

will not change much later on in the search. In other words, an inspector will either see or not see the defect fairly quickly, but the probability of detection will not change much with increased search time.

To illustrate the predictive property of the model, consider three different search durations: $t_1 = 3$s, $t_2 = 12$s, and $t_3 = 10$s. The mean search time is approximately $25/3 = 8 \sim 10$s. The cumulative probability function for this search can be modeled by the function $1 - e^{-t/10} = 1 - e^{-.1t}$, and is plotted in Figure 12.5. This particular instance of the visual search model would predict (roughly) that about 10 seconds are needed to achieve a 63% accuracy rate. A 95% success rate is reached at about 30 seconds.

The goal of modeling visual search in a visual inspection task is to be able to gauge an inspector's performance versus an idealized model of visual search such as the random search derived above. In contrast to this memory-less model, a perfect-memory systematic search is expected to reach a higher ac-

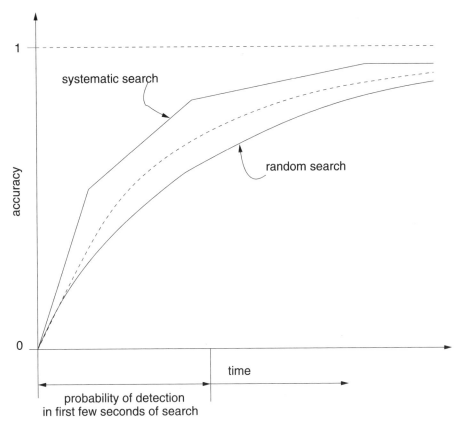

Fig. 12.4. Models of visual search.

curacy level faster than the random model. A systematic search model may be described by a piecewise-linear function as shown in Figure 12.4 (highest solid curve). The systematic search is simply another idealized model, one that happens to be very efficient in comparison with the random model. Human performance is expected to fall somewhere in between the two idealized models. That is, when a cumulative distribution of human visual search performance is plotted, it should appear somewhere between the two curves, as indicated by the dashed curve in Figure 12.4. Using this methodology, it is then possible to gauge the efficiency of an individual inspector–the closer the inspector's curve is to the idealized systematic model representation, the better the inspector's performance. This information can then be used to either rate the inspector, or perhaps to train the inspector to improve their performance.

Tracking eye movements during visual inspection may lead to similar predictive analyses, if certain recurring patterns or statistics can be found in collected

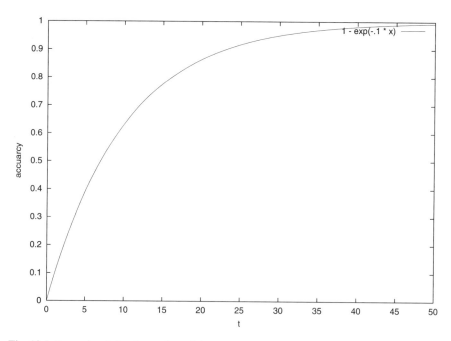

Fig. 12.5. Example of visual search model.

scanpaths. For example, an expert inspector's eye movements may clearly exhibit a systematic pattern. If so, then this pattern may be one that can be used to train novice inspectors. In the study of visual inspection of integrated circuit chips, Schoonard et al (1973) found that good inspectors are characterized by relatively high accuracy and relatively high speed, and make many brief eye fixations (as opposed to fewer longer ones) during the time they have to inspect.

In a survey of eye movements in industrial inspection, Megaw and Richardson (1979), identify the following relevant eye movement parameters:

- Fixation times. The authors refer to the importance of recording mean fixation times (perhaps fixation durations would be more appropriate). Megaw and Richardson state that for most inspection tasks, the expected average fixation duration is about 300ms. Longer fixation times are associated with the confirmation of the presence of a potential target and with tasks where the search times are short.
- Number of fixations. The number of fixations is a much more critical parameter in determining search times than fixation times and is sensitive to both task and individual variables.

- Spatial distribution of fixations. The coverage given to the stimulus material can be found by measuring the frequency of fixations falling in the elements of a grid superimposed over the display or by finding the frequency of fixations falling on specific features of the display. In both cases these frequencies correlate with the informativeness of the respective parts of the display as revealed by subjective estimates made by the searchers.
- Interfixation distances. With static displays this measure is equivalent to the amplitude of the saccadic movements. It is possible that when a comparatively systematic strategy is being employed, interfixation distances may reflect the size of the useful field of view (visual lobe).
- Direction of eye movements. Horizontal saccades may occur more frequently than vertical ones, which may reflect the elliptical shape of the effective useful field of view (visual lobe).
- Sequential indices. The most popular of these is the scanpath.

In their survey, the authors review previous inspection studies where eye movements were recorded. These include inspection of sheet metal, empty bottles, integrated circuits, and tapered roller bearings. In an inspection study of tin-plated cans, the authors report that experienced inspectors exhibited smaller numbers of fixations and that each inspector used the same basic scanpath from one trial to the next although there were small differences between inspectors. In an inspection study of electrical edge connectors, it was noted that the number of fixations per connector was much greater for complex items (i.e., more fixations are needed to search over a complex target).

Besides being useful for gauging inspection performance, eye movements may play a part in training visual search strategy (Wang, Lin, & Drury, 1997). Visual search strategy training can be effective in adoption of a desirable systematic search strategy. To train visual search strategy, eye movements may be used as both a feedback mechanism and as confirmation of adoption of the new search strategy. In their paper on search strategy training, Wang et al. recorded eye movements over visual inspection of artificial printed circuit boards. Eye movements were used to check how well each subject's search pattern followed the strategy trained. Scanpaths were judged by the experimenter after each training trial and feedback was given to the subject regarding whether they had followed instructions to search in a random or systematic manner.

Since training has been identified as the primary intervention strategy in improving inspection performance (see Gramopadhye, Bhagwat, Kimbler, and Greenstein (1998)), Duchowski et al (2001) investigated the utility of eye movements for search training in Virtual Reality. A three-dimensional aircraft inspection simulator was developed at Clemson University for this puropose. The aesthetic appearance of the environment is driven by standard graphical techniques augmented by realistic texture maps of the physical environment. The user's gaze direction, as well as head position and orientation, are tracked to allow recording of the user's fixations within the environment. The diagnostic eye tracking VR system allows recording of process measures (head and eye movements) as well as performance measures (search time and success rate) during immersion in the VR aircraft inspection simulator. The VR simulator features a binocular eye tracker, built into the system's Head Mounted Display (HMD), which allows the recording of the user's dynamic Point Of Regard (POR) within the virtual environment. A user wearing the HMD and an example of raw eye tracker output are shown in Figure 12.6. An experiment was conducted to measure the training effects of the VR aircraft inspection simulator. The objectives of the experiment included: (1) validation of performance measures used to gauge training effects, and (2) evaluation of the eye movement data as cognitive feedback for training. Assuming eye movement analysis correctly identifies fixations and the VR simulator is effective for training (i.e., a positive training effect can be measured), the number of detected fixations are expected to decrease with the adoption of an improved visual search strategy (Drury, Gramopadhye, & Sharit, 1997) (e.g., following training). The criterion task consisted of inspecting the simulated aircraft cargo bay in search of defects. Several defects can occur in a real environment situation. Three types of defects were selected to create inspection scenarios:

1. Corrosion: represented by a collection of gray and white globules on the inner walls of the aircraft cargo bay and located roughly at knee level.
2. Cracks: represented by a cut in any direction on the structural frames inside the aircraft cargo bay.
3. Damaged conduits: shown as either broken or delaminated electrical conduits in the aircraft cargo bay.

Data for performance and cognitive feedback measures was collected using search timing and eye movement information, respectively. The following performance measures were collected:

1. Search time from region presentation to fault detection.
2. Incremental stop time when subjects terminated the search in a region by deciding the region does not contain faults.
3. Number of faults detected (hits), recorded separately for each fault type.
4. Number of faults that were not identified (misses).

Fixation analysis enabled the collection of cognitive feedback measures, which were provided to subjects during the training session. Cognitive feedback measures were based on the eye movement parameters that contribute to search strategies as defined by Megaw and Richardson (1979), including: (1) total number of fixations; (2) mean fixation duration; (3) percentage area covered; and (4) total trial time. Cognitive feedback measures were graphically displayed off-line by rendering a 3D environment identical to the aircraft cargo bay which was used during immersive trials. This display represented the scanpaths of each trial to indicate the subject's visual search progression. An example scanpath is shown in Figure 12.7.

Analysis indicates that, overall, training in the VR aircraft simulator has a positive effect on subsequent search performance in VR, although there is apparently no difference in the type of feedback given to subjects. Cognitive feedback, in the form of visualized scanpaths, does not appear to be any more effective than performance feedback. It may be that the common most effective contributor to training is the immersion in the VR environment, that is, the exposure to the given task, or at least to the simulated task. Whether the eye tracker, by providing cognitive feedback, contributes to the improvement of inspection performance is inconclusive. Users may benefit just as much from performance feedback alone. However, the eye tracker is a valuable tool for collecting process measures. Analysis of results leads to two observations. First, mean fixation times do not appear to change significantly following training. This is not surprising since eye movements are to a large extent driven by physiology (i.e., muscular and neurological functions) and cognitive skill. In this case the search task itself may not have altered cognitive load per se, rather, prior experience in the simulator may have facilitated a more efficient search. Second, the number of fixations appears to decrease following training. These results generally appear to agree with the expectation of reduced number of fixations with the adoption of an improved visual search strategy (e.g., due to learning or familiarization of the task). The implication of reduced number of fixations (without an increase in mean fixation time) suggests that in the post-training case, subjects tend to employ a greater number of saccadic eye movements. That is, an improved visual search strategy may be one where subjects inspect the environment more quickly (perhaps due to familiarity gained through training), reducing the time required to visually rest on particular features.

Recently, Reingold, Charness, Pomplun, and Stampe (2002) review the motivation behind using chess as an ideal task environment for the study of skilled performance. Since pioneering work showing perception and memory to be more important differentiators of expertise than the ability to think ahead in the

(a) User wearing eye tracking HMD.

(b) Raw eye tracker output (left eye).

Fig. 12.6. Virtual aircraft inspection simulator.

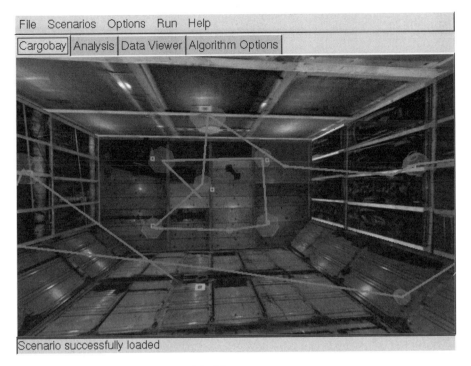

Fig. 12.7. Visualization of 3D scanpath in VR.

search for good moves, chess research has been instrumental in enhancing our understanding of human expertise and in contributing to the study of artificial intelligence. The chess master is thought to use recognizable configurations of pieces, chunks and templates as indices to long-term memory structures that, in association with a problem-solving context, trigger the generation of plausible moves for use by a search mechanism. Search is thereby constrained to the more promising branches in the space of possible moves from a given chess position. Hence, Grandmasters, the best human players, can find excellent moves despite generating only a small number of potential states (perhaps 100 or so for a few minutes of search). Such constrained search differs sharply from the enormous space explored by computer chess programs, which typically explore millions to 100s of millions of alternatives in the same time frame.

Reingold et al study visual span as a function of chess skill (expert vs. intermediate vs. novice) and configuration type (chess configuration vs. random configuration) using a gaze-contingent window technique. The paper extends classic work demonstrating that after viewing structured, but not random, chess positions for five seconds, chess masters reproduced these positions much more accurately than lesser skilled players. The authors document dramatically larger visual spans for experts while processing structured, but not random, chess positions. In a check detection task, where a minimized 3×3 chessboard containing a King and potentially checking pieces were displayed, experts made fewer fixations per trial, and had a greater proportion of fixations between individual pieces, rather than on pieces. These results provide strong evidence for a perceptual encoding advantage for experts attributable to chess experience, rather than to a general perceptual or memory superiority.

12.4 Summary and Further Reading

To summarize the potential impact of eye trackers in human factors research, this chapter focused on the eye tracker's capability of recording human visual process measures during exemplar tasks in aviation, driving, and during visual inspection. While performance measures typically quantify *how* a person performed (e.g., with what speed and accuracy), process measures can not only corroborate performance gains, but can also lead to discoveries of reasons for performance improvements (i.e., *what* the subject performed). In particular, tracking the users' eyes can potentially lead to further insights into the underlying human cognitive processes under varying conditions and workloads.

Good sources for further information dealing with eye tracking and human factors include the Human Factors and Ergonomics Society (HFES) journal and annual conference proceedings, and the proceedings of SIGCHI, the European Conference on Eye Movements (ECEM), and the US-based Eye Tracking Research & Applications (ETRA).

13. Marketing/Advertising

Eye tracking can aid in the assessment of ad effectiveness in such applications as copy testing in print, images, video, or graphics, and in disclosure research involving perception of fine print within print media and within available television and emerging High Definition TV (HDTV) displays.

A particularly illustrative although fictitious example of eye tracking in advertising can be seen in the movie *Looker* (Crichton, 1981). In one scene, the star of the movie, Dr. Larry Roberts, a cosmetic surgeon (played by Albert Finney), is shown a potential advertisement of a beauty product. The ad features an attractive model in a beach scene. The beauty product is displayed while Dr. Roberts' eyes are tracked. When it is obvious that Dr. Roberts' attention is drawn to the attractive model rather than the beauty product, the product is immediately moved to appear much closer to the point on the screen where Dr. Roberts' attention is drawn. While fictional, the effectiveness of eye tracking is effectively illustrated.

The motivation for utilizing an eye tracker in market research stems from the desire to understand consumer actions. In general, advertisers aim to provide product information to consumers in an efficient manner so that consumers' awareness of the existence of the product is heightened. If the consumer identifies the product as one which can potentially satisfy their current need, it is expected that the consumer will be more likely to purchase that particular product than if the consumer had not been aware of the product's availability. Consumer action can roughly be modeled by the block diagram shown in Figure 13.1. In general, a consumer's actions will be influenced by a combination of external and internal factors. External factors may include marketing actions (e.g., product promotion, distribution, availability), competitive factors (e.g., the desire to possess the latest and greatest product), and environmental factors (e.g., rainy conditions causing the need to possess an umbrella). Based on external influences and internal (perceptual and cognitive) processes, a consumer will make a choice as to whether or not to make a purchase, and which product to purchase, resulting in consumer action. The human decision making process may be affected by recognition of one's need or desire, and may also be influenced by external information gathered through research and/or through past memories and experiences.

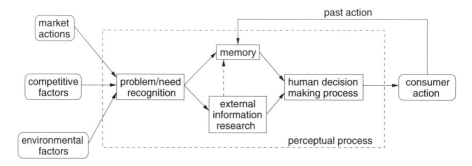

Fig. 13.1. Model of market and consumer actions.

In can be argued that external influences and resultant consumer actions are the only relevant and measurable factors to advertisers. That is, one can measure the relationship between marketing action (level of advertising) to the resulting consumer actions (sales of product). In this case, the model of the consumer may be treated as a black box: it is not as important to measure how the consumer functions, but rather just to measure what actions the consumer has performed (i.e., it may be sufficient to measure just the effect of consumer actions, not necessarily the cause). On the other hand, if one understands the cognitive and perceptual processes internal to the consumer, then a model of consumers' internal processes may aid the direction of marketing actions. A primary goal then is an understanding of the type of information that consumers want or use to make their decisions. Thus, if one can measure the perceptual process during the consumer's acquisition of information, it may be possible to tailor the information in a way such that the information the marketer wishes to impart to the consumer is delivered as efficiently and directly as possible. Eye tracking can provide insight into at least one aspect of the internal consumer model: how the consumer disperses visual attention over different forms of advertising. For example, current eye tracking technology can fairly easily provide a glimpse of the consumer's (overt) attentive processes over print media and television advertising, as illustrated above by the scene in the movie *Looker*. Indeed, it is quite plausible that market researchers are aware of eye trackers and utilize them for this purpose. Unfortunately, finding evidence of eye tracking in use by advertising companies is somewhat difficult. It appears that the use of eye trackers is not often well documented or otherwise advertised. Applied research organizations may routinely examine the eye movements of consumers as they look at advertisements, however, this work tends to be proprietary (Rayner, Rotello, Stewart, Keir, & Duffy, 2001). Here, only a few examples of published eye tracking work are presented, the rest of this chapter is augmented by examples of students' work and in-class projects.

13.1 Copy Testing

A particularly good example of analysis of eye movements over advertisements in the Yellow Pages™ is given by Lohse (1997). In this experiment, eye movement data is collected while consumers chose businesses from telephone directories. The study addresses (1) what particular features cause people to notice an ad, (2) whether people view ads in any particular order, and (3) how viewing time varies as a function of particular ad features. Based on a review of literature on attention and a brief survey of prototypical eye movement patterns, the author develops the following propositions:

1. Color: Yellow Pages™ users are more likely to notice color ads before any other type of ad.
2. Graphics: Yellow Pages™ users are likely to notice ads with graphics before ads without graphics.
3. Size: Yellow Pages™ users are likely to notice large ads before small ads.
4. Location: Yellow Pages™ users are more likely to view advertisements near the beginning of the heading than those near the end of the heading.

It is interesting to note some important points of the experimental methodology. To prevent bias toward any recognizable businesses, a completely artificial mockup of a 32-page Yellow Pages™ directory was created. The pages were virtually indistinguishable from real Yellow Pages™ in terms of font, ink, and color, but all business entries were fabricated. The 32 directory pages were organized into four books to control for combinations of the following layout and design features: ad type, location of display ads on the page, size of ad, color, use of graphics, whether or not a listing had a bold typeface, serial position of the ad (alphabetic order), and number of types of information in the ad (hours, years in business, slogan, brand names, specialties). Eye movement analysis revealed that the results from the Yellow Pages™ study are consistent with previous findings on print advertising in magazines, catalogs, and newspapers. Ad size, graphics, color, and copy all influence attention to advertisements. The author offers the following observations:

1. Color and graphics: color ads with graphics capture attention. Color ads were scanned more quickly, more often, and longer than black and white ads. Subjects noticed more color ads than ads without color (92% vs. 84%) and viewed color ads before ads without color. Subjects viewed color ads 21% longer than equivalent ads without color. Subjects also viewed 96% of ads with graphics. However, unlike color, graphics did not capture initial consumer attention.

2. Size: ad size influenced attention. In general, the larger the ad, the more likely subjects were to notice the ad. Subjects noticed 93% of the large display ads but only 26% of the plain listings. Quarter-page ad displays were much more noticed than text listings.

3. Location: the position of an ad on the page had a large effect on whether people viewed the ad, even though the position says nothing informative about the business. Position matters because people scan ads on a page in alphabetic order and their scan is not exhaustive; as a result, people never read some ads.

Due to the improvement of eye tracking technology (e.g., bite bars are no longer needed), the author suggests that the time has come to reevaluate the importance of eye tracking equipment as a tool for print advertising research.

13.2 Print Advertising

In a study of consumers' visual attention over print advertisements, an eye tracker was used to gain insight into attentive processes over repeated exposure to print advertisements (Rosbergen, Wedel, & Pieters, 1990). The authors explore the phenomenon of repeated advertising's "wearout", i.e., the authors investigate consumers' diminishing attentional devotion to ads with increased repetition. Consumers' visual attention is measured to key print ad elements: headline, pictorial, bodytext, and packshot. A statistical model is proposed comprising submodels for three key measures of visual attention to specific elements of the advertisement: attention onset, attention duration, and inter- and intra-element saccade frequencies. Analyses show that whereas duration decreases and attention onset accelerates during each additional exposure to the print ad, the attentional scanpath remains constant across advertising repetitions and across experimentally varied conditions. The authors also list their important findings:

- Attention durations differ significantly across ad elements. Parameter estimates reveal that attention duration is longest for the text, followed by headline, and shortest for the pictorial and the packshot. In addition, a progressive decrease in the expected attention duration is observed across exposures. Neither motivation nor argument quality affects the amount of attention paid to ad elements.

- The time until subjects first fixate on an ad element differs among elements. Subjects attend first to the headline followed by the pictorial, the text, and finally the packshot. Although repetition as such has no impact on attention onset, differences in attention onset are not constant across exposures. Less time lies between the expected starts of the first fixations during the second and third exposures than during the first exposure. In other words, the attentional process accelerates during later exposures.

Attention onset is also significantly affected by motivation in the sense that attention onsets are farther apart for highly motivated subjects than for less motivated subjects. However, motivation does not change the order in which ad elements are attended to for the first time nor does the impact of motivation on attention onset differ across exposures.

- Analysis and modeling of scanpaths suggest that the ad's scanpath can be described by a reversible, stationary first-order Markov process. Based on this model, expected transition matrices suggest that:
  1. The amount of attention paid to the text is about three times as high as the amount paid to the pictorial.
  2. The amount of attention paid to the ad decreases by about 50% from exposure 1 to exposure 3.
  3. The majority of saccades (about 75%) occur within ad elements, in particular in the bodytext.
  4. Most inter-element saccades start from or end at the packshot.
  5. The expected transition matrices are quasi-symmetric.
  6. The conditional transition probabilities remain constant across exposures.

Combining the results with the authors' modeling efforts yields the following observations. As the attention onsets show, subjects attend, on average, first to the headline. As indicated by the expected number of saccades between headline and pictorial, attention is then directed to the pictorial. However, attention onsets provide some indication that during later exposures this order may be reversed. Both headline and pictorial receive about one-sixth of subjects' attention. Half of the attention is directed at the bodytext, but subjects focus on the bodytext only after the headline and the pictorial have received some initial attention. Finally, subjects attend to the packshot last, and that, despite the limited amount of attention spent on this ad element, most intra-element saccades start from and end at the packshot. This may point to integration of information in other ad elements with information in the packshot.

In a recent study of eye movements over advertisements, Wedel and Pieters (2000) comment that while eye movements are eminent indicators of attention, what is currently missing in eye movement research is a serious account of the processing that takes place to store information in long-term memory. The authors attempt to provide such an account through the development of a formal model. They model the process by which eye fixations on print advertisements lead to memory for the advertised brands. The model is calibrated to eye movement data collected during exposure of subjects to magazine ads and subsequent recognition of the brand in a perceptual memory task. Available data for each subject consist of the frequency of fixations on the ad elements (brand, pictorial, and text), and the accuracy and the latency of memory. It is assumed that the number of fixations, not their duration, is related to the amount of information a consumer extracts from an ad. The accumulation of information across multiple fixations to the ad elements in long-term memory is assumed to be additive. Wedel and Pieters's model is applied in a study involving a sample of 88 consumers who were exposed to 65 print ads appearing in their natural context in two magazines. Wedel and Pieters report that across the two magazines, fixations to the pictorial and the brand systematically promote accurate brand memory, but text fixations do not. Brand surface has a particularly prominent effect. The more information is extracted from an ad during fixations, the shorter the latency of brand memory is. A systematic recency effect was found: when subjects are exposed to an ad later, they tend to identify it better. In addition, there is a small primacy effect. The effect of the ad's location on the right or left of the page depends on the advertising context.

Considering text and pictorial information in advertisement, Rayner et al (2001) performed a study where viewers looked at print advertisements as their eye movements were recorded. Eye movements were recorded by an EyeLink headband-mounted tracker from SensoMotoric Instruments (SMI). Although viewing was binocular, movements of the right eye were monitored. Eye positions were sampled at 250 Hz. Half the viewers were told to pay special attention to car ads, and the other half were told to pay special attention to skin-care ads. Rayner et al found that viewers tended to spend more time looking at the text than the picture part of the ad, though they did spend more time looking at the type of ad they were instructed to pay attention to. Fixation durations and saccade lengths were both longer on the picture part of the ad than the text, but more fixations were made on the text regions. Viewers did not alternate fixations between the text and picture part of the ad, but they tended to read the large print, then the smaller print, and then they looked at the picture (although some viewers did an initial cursory scan of the picture). Of the 110 product names that were generated in the free-recall task, 89 (81%) were recalled correctly. Overall recognition performance was excellent. Despite participants' focus on the text in the ads, memory for the product names was not particularly good. On average, participants correctly recalled fewer than four brand names from the 24 advertisements they had studied. In addition, the advertisements that were viewed favorably were often not identified by brand name; those preferred ads were often described to the experimenter in terms of some aspect of the pictorial information or by a generic product label.

Although Rayner et al report that scan path data from the experiment are difficult to quantify, they note some very striking characteristics of how viewers scanned the ads. The initial fixation on the ad was always located in the center of the ad because that is where the viewer was fixated when the ad was initially presented. Looking behavior was fairly consistent across viewers in that they typically initially made an eye movement to the large print, regardless of its spatial position within the ad. After looking at the large print, participants either made a very cursory scan of the picture, or more typically, they moved from the large print to the small print and then to the picture.

Rayner et al suggest that the presented data have some striking implications for applied research and advertisement development. Instructions given to participants can influence their eye movements suggests that participants' goals must be considered in future research conducted by advertising agencies and researchers in the area. The authors note that the data also indicates that an advertisement captures and holds participants' attention but this may be caused by the instructions given to those participants. If different instructions were used, the evidence of the advertisements' success in capturing attention may be reduced. The authors note that their second main finding, that more time was spent viewing text than viewing pictures in these ads, is inconsistent with how advertising agencies view the relative importance of the visual and text portions of the ads. Early research on print advertisements concluded that an ad is typically well liked during prepublication testing if it has a single illustration, a short headline, lots of white space and very little text. Data presented here indicate that consumers may be paying much more attention to the text in ads than previously thought. Rayner et al conclude that currently there is no cognitive process theory that makes clear predictions about eye movements and attention while viewing print advertising. The data and ideas outlined in this paper have the potential to form the foundation in the search for such a theory.

13.3 Ad Placement

Eye movements recorded over advertisements are particularly informative since scanpaths immediately provide a visual depiction of whether the intended text or object was fixated (or at least scanned over). This was a topic of study of a pair of senior undergraduate students, one from Computer Science, the other from Marketing. Examples of scanpaths recorded over advertisement images are shown in Figure 13.2. While formal analyses cannot be offered, the images show fairly typical scanpaths over ads. It can be seen that attention is drawn to fairly conspicuous ad elements, such as faces, textual information, and objects set apart by virtue of being presented in homogeneous color regions.

Another Marketing student study was conducted over NASCAR™ (National Association for Stock Car Auto Racing) images. The student team consisted of majors from Computer Science, Marketing, and Industrial Engineering. The distribution of work fell along fairly well expected lines of specialization: the Computer Science major provided coding support, the Industrial Engineer provided experimental design and statistical analysis expertise, and the Marketing students provided contextual and background information (motivation for the study, image samples, etc.). The objective of the experiment was simply to find whether any particular regions on the vehicle tend to inherently draw attention. Images from the study are given in Figure 13.3.

Different views of NASCAR™ vehicles were evaluated: a front-right view (as shown in Figure 13.3), a front-left view, and left and right views of the vehicle rears. These images were chosen to represent views of vehicles as may be seen during a televised NASCAR™ event. Specific vehicle regions were chosen a priori as Regions Of Interest (ROIs): the hood, middle, rear quarter panel, and trunk, depending on the view of the vehicle. Pictures viewed by subjects were grouped based on the visibility of ROIs. Mean viewing times and number of fixations were compared between groups. The control stimulus was an image of a NASCAR™ vehicle with the ads airbrushed out (as shown in Figure 13.3(b)). The control stimulus was used as an image meant to be unbiased by advertising content. Informal analysis based on the number of fixations within ROIs suggests that the shape of the vehicle alone is responsible for drawing attention rather than the ads placed themselves. From images showing the left and right rear views of the vehicles, it was determined that the

(a) Altoids ad.

(b) M&Ms ad.

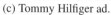

(c) Tommy Hilfiger ad.

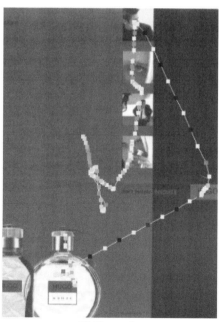

(d) Hugo ad.

Fig. 13.2. Scanpaths over advertisements. Courtesy of Cristy Lander and Karen Kopp. Reproduced with permission, Clemson University.

(a) Front-right view of vehicle.   (b) Advertisements air-brushed out.

(c) Identified fixations.   (d) Highlighted Regions Of Interest (ROIs).

Fig. 13.3. Scanpaths over NASCAR™ vehicles. Courtesy of Melissa Andrews, Laura Boyd, Robyn Bushee, and Amit Joshi. Reproduced with permission, Clemson University.

rear quarter panel draws most of the viewer's attention (the highest number of fixations were counted in this ROI). From images showing the right and front views of the vehicles, the middle section was found to draw most of the viewers' attention.

Due to the short, in-class nature of these studies, significance and generalizability of results can only be claimed to be anecdotal. Results reported here are not meant to be conclusive or significant. However, the exercises are excellent examples of potential student-led eye tracking projects in testing advertisement materials.

13.4 Summary and Further Reading

There are numerous opportunities for conducting eye tracking Marketing studies. Copy testing, print advertising, and ad placement are suitable potential experiments that may be used to improve the impact of advertising materials. Unfortunately, evidence of eye tracking in market research is difficult to find.

It may be that advertising companies do not wish to disclose the fact that eye trackers are being used. This may be perceived by the buying public as somehow being devious. Still, it is fairly safe to say that eye trackers are probably well known to marketing researchers and with improvements in technology will continue to be valuable tools in their work.

Possible sources of eye tracking research include scientific journals and professional conferences. One particular source that occasionally contains reports of eye tracking work is the *Journal of Advertising*. Another place to search for evidence of eye tracker use is the World Wide Web. A quick search of the web reveals the following available reports:

- The Outdoor Advertising Association of America (`<http://www.oaaa.org/>`) lists the following available research report: "The PRS Eye Tracking Studies: Validating Outdoor's Impact in the Marketplace" (last accessed 05/31/02).
- The Institute of Behavioural Sciences `<http://ibs.derby.ac.uk/>` offers its eye tracking services for examining printed material (see `<http://ibs.derby.ac.uk/research/eye.html>`, last accessed 01/03/02).
- A conference announcement by the Advertising Research Foundation (ARF), Week Of Workshops (WOW), October 29-November 1, 2001, lists among its scheduled presentations the following: "New Ad Designs Capture Users' Eyes: A Case Study of Eye Tracking for CNET". (See `<http://www.thearf.org/webpages/wow-2001/conference-tuesday-marketing.html>`, last accessed 01/03/02.)

Thus it appears eye tracking market research is being conducted. The last example is particularly interesting since it shows a fairly new domain ripe for the study of advertising placement: the world wide web itself.

14. Computer Science

This chapter gives an overview of primarily interactive applications mainly developed by computer science researchers. Following the hierarchy of eye tracking systems given earlier, apart from diagnostic usability studies, this chapter focuses on two types of interactive applications: selective and gaze-contingent. The former approach uses an eye tracker as an input device, similar in some ways to a mouse. This type of ocular interaction is often studied by researchers involved in the fields of Human-Computer Interaction (HCI) and Computer-Supported Collaborative Work (CSCW). The latter gaze-contingent application is a type of display system wherein the information presented to the viewer is generally manipulated to match the processing capability of the Human Visual System, often matching foveo-peripheral perception in real-time. It should be noted that a good deal of previous work discussed in this chapter is based on conference proceedings.

14.1 Human-Computer Interaction and Collaborative Systems

Eye-based interactive systems have been presented at several SIGCHI conferences with a significant increase in the number of papers in recent years. Most of these papers have traditionally focused on interactive uses of eye trackers, although diagnostic applications have also begun to appear, particularly in the context of usability studies.

Interactive uses of eye trackers typically employ gaze as a pointing modality, e.g., using gaze in a similar manner to a mouse pointer. Prominent applications involve selection of interface items (menus, buttons, etc.), as well as selection of objects or areas in Virtual Reality (VR). A prototypical "real-world" application of gaze as an interactive modality is eye typing, particularly for handicapped users.

Other uses of gaze in the general field of Human-Computer Interaction involve gaze as an indirect pointing modality, for example as a deictic reference in the context of collaborative systems, or as an indirect pointing aid in user interfaces. Diagnostic uses of eye trackers are becoming adopted for usability studies, i.e., testing the effectiveness of interfaces as evidenced by where users look on the display.

14.1.1 Eye-Based Interaction

One of the first eye-based interactive systems, introduced by Jacob (1990), demonstrated an intelligent gaze-based informational display. In Jacob's system a text window would scroll to show information on visually selected items. The paper's title, "What You Look At Is What You Get", is a play on words on the common word processing/printing paradigm of What You See Is What You Get (WYSIWYG). Jacob's paper was an early paper describing the feasibility of a gaze-based interactive system in which the author discussed the possibility of using gaze in place of or in addition to a mouse pointer. The paper was one of the first to use video-based corneal reflection eye tracking technology interactively and is well known for its identification of an important problem in eye-based interactive systems: the Midas Touch problem. Essentially, if the eyes are used in a manner similar to a mouse, a difficulty arises in determining intended activation of foveated features (the eyes do not register button clicks!). That is, unlike a mouse with which a user signifies activation of an object by pressing a mouse button, with gaze pointing everything that a user looks at is potentially activated (and so in the Midas analogy unintentionally turns to gold). To avoid the Midas Touch problem, Jacob discusses several possible solutions including blinks, finally promoting the use of dwell time (of about 150-200ms) to act as a selection mechanism.

At the same SIGCHI meeting as in which Jacob's paper appeared, a graphical "self-disclosing" display was presented by Starker and Bolt (1990). This interactive system provided the user with gaze-controlled navigation in a three-dimensional graphics world. The graphical environment contained story world characters who responded in interesting ways to the user's gaze. Fixations activated object "behaviors", since the system would maintain and increase the user's visual interest level in an object. With increased interest, objects would blush and/or provide a verbal narrative. Unlike Jacob's use of dwell time, in this system dwell time was used to zoom into the graphics world.

At a more recent SIGCHI meeting, Tanriverdi and Jacob (2000) presented a new gaze-based interactive system, this time with gaze acting as a selective mechanism in VR. In this system Tanriverdi and Jacob compared eye-based interaction with hand-based interaction. The authors found that performance with gaze selection was significantly faster than hand pointing, especially in environments where objects were placed far away from the user's location. In contrast, no performance difference was found in "close" environments. Furthermore, while pointing speed may increase with gaze selection, there appears to be a cognitive tradeoff for this gain in efficiency: subjects had more difficulty recalling locations they interacted with when using gaze-based selection than when using hand selection.

Besides pointing in desktop and immersive (VR) displays, the archetypical gaze-based pointing application is eye typing. Eye typing is and has been a useful communication modality for the severely handicapped. Severely handicapped people, possessing an acute need for a communication system, may only be able to control their eyes. Most eye typing systems are implemented by presenting the user with a virtual keyboard, either on a typical computer monitor, or in some cases projected onto a wall. Based on analysis of tracked gaze, the system decides which letter the user is looking at and decides (e.g., by dwell time) whether or not to type this letter. The system may provide feedback to the user by either visual or auditory means, or a combination of both. Eye tracking systems may be either video-based or based on electro-oculographic (EOG) potential. An example eye typing interface is shown in Figure 14.1. Variations of traditional gaze-based pointing may be employed for subjects exhibiting difficulties fixating (e.g., locked-in syndrome). In such cases, if the eyes can only be moved in one direction, the eyes can be used as simple one- or two-way switches and the focus can be shifted from one item to another by using a method known as scanning. Using a combination of scanning and switching, the user can use one switch to change display focus by scanning across the display, then use another switch to indicate selection of the item currently in focus. Majaranta and Raiha (2002) provide an excellent survey of additional selection and feedback techniques employed in eye typing systems.

Gaze-based communication systems such as those featuring eye typing offer certain (sometimes obvious) advantages but are also problematic. Gaze may provide an often faster pointing modality than a mouse or other pointing device, especially if the targets are sufficiently large. However, gaze is not as accurate as a mouse since the fovea limits the accuracy of the measured point of regard to about 0.5 degrees visual angle. Another significant problem is accuracy of the eye tracker. Following initial calibration, eye tracker accuracy may exhibit significant drift, where the measured point of regard gradually falls off

Fig. 14.1. Example of eye typing interface. Courtesy of Prentke Romich Company, 1022 Heyl Road, Wooster, OH 44691 USA <http://www.prentrom.com/access/-hm2000.html>. Reproduced with permission.

from the actual point of gaze. Together with the Midas Touch problem, drift remains a significant problem for gaze input.

Zhai, Morimoto, and Ihde (1999) take another approach to gaze-based interaction and test the use of gaze as a sort of predictive pointing aid rather than a direct effector of selection. This is a particularly interesting and significant departure from "eye pointing" since this strategy is based on the authors' assertion that loading of the visual perception channel with a motor control task seems fundamentally at odds with users' natural mental model in which the eye searches for, and takes in information, while coordinating with the hand for manipulation of external objects. In their paper on Manual Gaze Input Cascaded (MAGIC) Pointing, Zhai et al present an acceleration technique where a (two-dimensional) cursor is warped to the vicinity of a fixated target. The acceleration is either immediate, tied to eye movement (liberal MAGIC pointing), or delayed, following the onset of mouse movement (conservative MAGIC pointing). The authors report that while the speed advantage is not obvious over manual (mouse) pointing (subjects tended to perform faster with the liberal method), almost all users subjectively felt faster with either new pointing technique.

14.1.2 Usability

Besides the use of gaze for interactive means, diagnostic eye tracking is gaining acceptance within the HCI community as another means to test usability of an interface. It is believed that eye movements can significantly enhance the observation of users' strategies while using computer interfaces (Goldberg & Kotval, 1999). Among various experiments, eye movements have been used to evaluate the grouping of tool icons, compare gaze-based and mouse interaction techniques (Sibert & Jacob, 2000), evaluate the organization of click-down menus, and more recently to test the organization of web pages.

In their paper presented at SIGCHI, Byrne, Anderson, Douglass, and Matessa (1999) tested the arrangement of items during visual search of click-down menus. The authors contrast two computational cognitive models designed to predict latency, accuracy, and ease of learning for a wide variety of HCI-related tasks: EPIC (Executive Process Interactive Control) (Kieras & Meyer, 1995) and Adaptive Control of Thought-Rational (ACT-R) (Anderson, 2002). The EPIC architecture provides a general framework for simulating a human interacting with their environment to accomplish a task (Hornof & Kieras, 1997). ACT-R is a framework for understanding human cognition whose basic claim is that cognitive skill is composed of production rules (Anderson, 1993). These models specifically predict, in different ways, the relationship between eye and mouse movement. With respect to click-down menus, the EPIC model predicts the following:

1. The distance covered by saccades is constant.
2. Eyes tend to overshoot the target with some regularity.
3. No mouse movement is initiated until the target is visually acquired.
4. Eye movement patterns are generally top-down and random.

In contrast, the ACT-R model predicts:

1. The distance covered by saccades is variable.
2. Eyes never overshoot the target.
3. Mouse movement generally follows saccades.
4. Eye movement patterns are exclusively top-down.

A visual search experiment was conducted where a target item was presented prior to display of a click-down menu. Based on the number of fixations, both models were supported since in some cases visual search exhibited exclusively top-down search (ACT-R) and in others both top-down and random patterns were observed (EPIC). Fixation positions, however, did not appear random, they did not appear to be strictly linear, and there appeared to be a preference for the first item in the menu. This particular study is informative for two reasons. First, it shows the importance of a good model (or lack of one) which can be used for designing user interfaces. If a model can be developed which successfully predicts some aspect of user activity, then this model may be used to design an interface which benefits the user by adapting to the user (instead of, for example, forcing the user to learn a new, possibly unintuitive style of interaction). The idea of modeling human behavior is an important concept in Human Computer Interaction. Second, the paper points out that eye tracking is an effective usability modality and that a new model for visual search over menus is needed, e.g., possibly a "noisy" version of top-down search.

In a usability study of web pages, Goldberg, Stimson, Lewnstein, Scott, and Wichansky (2002) derive specific recommendations for a prototype web interface tool. The authors discuss gaze-based evaluation of web pages in which the system permits free navigation across multiple web pages. This is a significant advancement for usability studies of web browsers since prior to this study recording of gaze over multiple web pages has been difficult due to the synchronization of gaze over windows that scroll or hide from view. The authors describe their collection of the following dependent measures:

1. User actions such as key presses and mouse button clicks.
2. Context-free eye movement measures such as fixations and dwell times.
3. Context-sensitive eye movements such as dwell times within regions of interest.

The key questions examined are whether eye movements are related to user actions, is navigation biased toward horizontal or vertical navigation, and whether there is any particular order to web page scanning. Following task-level, screen-level, and object-level analyses, the authors report that users exhibit a preference for horizontal search across columns rather than searching within a column. Furthermore, recommendations are made for the left-hand placement of web "portlets" requiring visibility, with the two most important portlets placed on top of the web page.

A good deal many more gaze-based usability studies have been conducted. The above are but two examples. A workshop organized by Karn et al (1999) at the SIGCHI 99 meeting drew a fairly large number of participants. It is expected that eye tracking will continue to play an increasingly significant role in usability investigations.

14.1.3 Collaborative Systems

Apart from interactive or diagnostic uses of eye movements, gaze can also be utilized to aid multiparty communication in collaborative systems. In the GAZE Groupware system, an LC Technologies eye tracker is used to convey gaze direction in a multiparty teleconferencing and document sharing system, providing a solution to two problems in multiparty mediated communication and collaboration: knowing who is talking to whom, and who is talking about what (Vertegaal, 1999). The system displays 2D images of remotely located participants in a VRML virtual world. These images rotate to depict gaze direction alleviating the problem of turn-taking in multiparty communication systems. Furthermore, a gaze-directed "lightspot" is shown over a shared document indicating the users' fixated regions and thereby providing a deictic ("look at this") reference. The system display is shown in Figure 14.2, with the the system interface shown in Figure 14.3. For further information, see: <http://www.cs.queensu.ca/home/roel/gaze/home.html>.

Fig. 14.2. GAZE Groupware display. Courtesy of Roel Vertegaal.

14.2 Gaze-Contingent Displays

In general, eye-based interactive applications can be thought of as selective, since gaze is used to select or point to some aspect of the display, whether it is two-dimensional (e.g., desktop), collaborative, or immersive (such as a virtual environment). Mixing both directly interactive and indirectly "passive" usage styles of gaze are gaze-contingent displays. Here, gaze is used not so much as a pointing device, but rather as a passive indicator of gaze. Given the user's point of regard, a system can tailor the display so that the most informative

(a) User interface.

(b) Eye tracking optics.

Fig. 14.3. GAZE Groupware interface. Courtesy of Roel Vertegaal.

details of the display are generated at the point of gaze, and degraded in some way in the periphery. The purpose of these displays is usually to minimize bandwidth requirements, as in video telephony applications, or in graphical applications where complex data sets cannot be fully displayed in real-time. Two main types of gaze-contingent applications are discussed: screen-based and model-based. The former deals with image (pixel) manipulation, while the latter is concerned with the manipulation of graphical objects or models. Both systems are generally investigated by researchers studying Computer Graphics (CG) and Virtual Reality (VR).

Human visual perception of digital imagery is an important contributing factor to the design of perceptually-based image and video display systems. Human observers have been used in various facets of digital display design, ranging from estimation of corrective display functions (e.g., gamma function) dependent on models of human color and luminance perception, color spaces (e.g., CIE Lab color space), and image and video codecs. JPEG and MPEG both use quantization tables based on the notion of Just Perceptible Differences to quantize colors of perceptually similar hue (Wallace, 1991).

The idea of gaze-contingent displays is not new and dates back to early military applications (Kocian, 1987; Longridge et al, 1989). In the Super Cockpit Visual World Subsystem, Kocian considered visual factors including contrast, resolution, and color in the design of a head-tracked display. In their Simulator Complexity Testbed (SCTB), Longridge et al included an eye-slaved ROI as a major component of the Helmet Mounted Fiber Optic Display (HMFOD). This ROI provided a high resolution inset in a low resolution (presumably homogeneous) field which followed the user's gaze. The precise method of peripheral degradation was not described apart from the criteria of low resolution. However, the authors did point out that a smooth transition between the ROI and background was necessary in order to circumvent the possibility of a perceptually disruptive edge artifact.

Various gaze-contingent approaches have been proposed for foveal Region Of Interest (ROI)-based image and video coding (Stelmach & Tam, 1994; Nguyen, Labit, & Odobez, 1994; Kortum & Geisler, 1996; Tsumura, Endo, Haneishi, & Miyake, 1996). Often, however, these studies are based on automatically located image regions, which may or may not correspond to foveally viewed segments of the scene. That is, these studies do not necessarily employ an eye tracker to verify the ROI-based coding schemes. Instead, a figure-ground assumption is often used to argue for more or less obvious foveal candidates in the scene. This is a research area where either diagnostic eye movement studies can be used to corroborate the figure-ground assumption, or gaze may be used directly to display high-resolution ROIs at the point of regard in real-time (as in eye-based teleconferencing systems).

Instead of assuming a feature-based approach to foveal (high-resolution) encoding, an eye tracker can be used to directly establish the foveal ROI and a suitable image degradation scheme may be employed to render detail at the point of regard. This motivated research into finding a suitable image degradation scheme which would match foveal acuity (Duchowski, 2000). Using the Discrete Wavelet Transform for smooth image resolution degradation, images demonstrating three acuity mapping functions are shown in Figure 14.4. For demonstration purposes, the *cnn* image was processed with an artificially placed ROI over the anchor's right eye and another over the "timebox" found in the bottom right corner of the image. Haar wavelets were used to accentuate the visibility of resolution bands (see Section 4.2). Figure 14.4(b), (d), and (f) show the extent of wavelet coefficient scaling in frequency space. The middle row shows a reconstructed image where resolution drops off smoothly, matching visual acuity for a particular screen display at a particular viewing distance.

14.2.1 Screen-Based Displays

When evaluating Gaze-Contingent Displays (GCDs), it is often necessary to distinguish between to main types of causalities: those affecting perception and those affecting performance. As a general rule, perception is more sensitive than performance. That is, it may be possible to degrade a display to a quite noticeable effect without necessarily degrading performance. In either case, one of the main difficulties that must be addressed is the latency of the system.

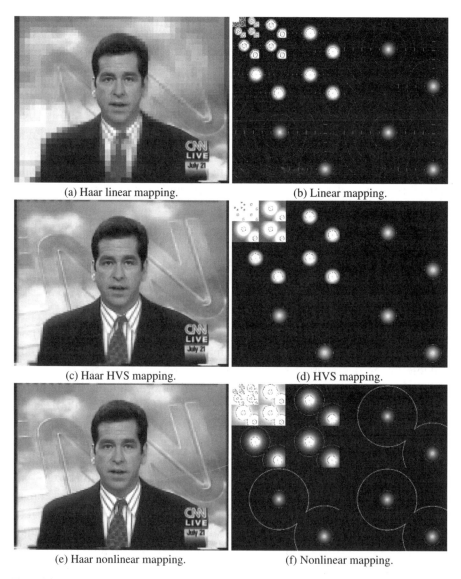

(a) Haar linear mapping. (b) Linear mapping.

(c) Haar HVS mapping. (d) HVS mapping.

(e) Haar nonlinear mapping. (f) Nonlinear mapping.

Fig. 14.4. Image reconstruction and wavelet coefficient resolution mapping (assuming 50dpi screen resolution). Reprinted from Duchowski (2000) with permission © 2000 IEEE.

Without predictive capabilities, most gaze-contingent displays will lag behind the user's gaze somewhat, usually by a constant amount of time proportional to both measurement of gaze (which may take up to the time it takes to display a single frame of video, typically 16ms for a 60Hz video-based tracker), and the subsequent time it takes to refresh the gaze-contingent display (this may take another 33ms for a system with update rate of 30 frames per second).

Loschky and McConkie (2000) conducted an experiment on a gaze-contingent display investigating spatial, resolutional, and temporal parameters affecting perception and performance. Two key issues addressed by the authors are the timing of GCDs and the detectability of the peripherally degraded component of the GCD. That is, how soon after the end of an eye movement does the window need to be updated in order to avoid disrupting processing, and is there a difference between the window sizes and peripheral degradation levels that are visually detectable and those that produce behavioral effects? In all experiments, monochromatic photographic scenes were used as stimuli with a circular, high-resolution window surrounded by a degraded peripheral region. An example of Loschky and McConkie's GCD is shown in Figure 14.5(a). In one facet of the experiment, it was found that for an image change to go undetected, it must be started within 5 ms after the end of the eye movement. Detection likelihood rose quickly beyond that point. In another facet of the study concerning detection of peripheral degradation, results showed that the least peripheral degradation (inclusion of 4 of 4 possible levels) went undetected even at the smallest window size (2°), where the opposite was true with the highest level of degradation–it was quite detectable at even the largest window size (5°). The GCD was also evaluated in terms of performance effects, in the context of visual search and scene recall tasks. In the end it was found that the generation of an imperceptible GCD was quite difficult in comparison to the generation of a GCD which does not deteriorate performance. While greater delays (e.g., 15ms) and greater degradation (inclusion of only 3 of 4 possible levels) produce detectable visual artifacts, they appear to have minimal impact on performance of visual tasks when there is a 4.1° high-resolution area centered at the point of gaze.

Parkhurst et al (2000) investigated behavioral effects of a two-region gaze-contingent display. A central high-resolution region, varying from 1 to 15 degrees, was presented at the instantaneous center of gaze during a visual search task. An example of Parkhurst et al.'s display is shown in Figure 14.5(b). Measures of reaction time, accuracy, and fixation durations were obtained during a visual search task. The authors' primary finding is that reaction time and accuracy co-vary as a function of the central region size. The authors note this as a clear indicator of a strategic speed/accuracy tradeoff where participants favor speed in some conditions and accuracy in others. For small central region sizes, slow reaction times are accompanied by high accuracy. Conversely, for large

(a) From Loschky and McConkie (2000) © 2000 ACM, Inc. Reprinted by permission.

(b) From Parkhurst, Culurciello, and Niebur (2000) © 2000 ACM, Inc. Reprinted by permission.

Fig. 14.5. Example gaze-contingent displays.

central regions sizes, fast reaction times are accompanied by low accuracy. A secondary finding indicated that fixation duration varies as a function of central region size. For small central region sizes, participants tend to spend more time examining each fixation than under normal viewing conditions. For large central regions, fixation durations tend to be closer to normal. In agreement with reaction time and accuracy, fixation duration is approximately normal (comparable to that seen for uniform resolution displays) with a central region size of $5°$.

For screen-based VR rendering the work of Watson et al (1997) is particularly relevant. The authors studied the effects of Level Of Detail (LOD) peripheral degradation on visual search performance. Both spatial and chrominance detail degradation effects were evaluated in Head Mounted Displays (HMDs). To sustain acceptable frame rates, two polygons were texture mapped in real-time to generate a high resolution inset within a low resolution display field. The authors suggested that visual spatial and chrominance complexity can be reduced by almost half without degrading performance.

In an approach similar to Watson et al's, Reddy (1998) used a view-dependent LOD technique to evaluate both perceptual effects and system performance gains. The author reported a perceptually modulated LOD system which affords a factor 4.5 improvement in frame rate.

14.2.2 Model-Based Graphical Displays

As an alternative to the screen-based peripheral degradation approach, model-based methods aim at reducing resolution by directly manipulating the model geometry prior to rendering. The technique of simplifying the resolution of geometric objects as they recede from the viewer, as originally proposed by Clarke (1976), is now standard practice, particularly in real-time applications such as VR (Vince, 1995). Clarke's original criteria of using the projected area covered by the object for descending the object's LOD hierarchy is still widely used today. However, as Clarke suggested, the LOD management typically employed by these polygonal simplification schemes relies on pre-computed fine-to-coarse hierarchies of an object. This leads to uniform, or *isotropic*, object resolution degradation.

A gaze-contingent model-based adaptive rendering scheme was proposed by Ohshima, Yamamoto, and Tamura (1996), where three visual characteristics were considered: central/peripheral vision, kinetic vision, and fusional vision. The LOD algorithm generated isotropically degraded objects at different visual angles. Although the use of a binocular eye tracker was proposed, the system as discussed used only head tracking as a substitute for gaze tracking.

Isotropic object degradation is not always desirable, especially when viewing large objects at close distances. In this case, traditional LOD schemes will display an LOD mesh at its full resolution even though the mesh may cover the entire field of view. Since acute resolvability of human vision is limited to the foveal 5°, object resolution need not be uniform. This is the central tenet of gaze-contingent systems.

Numerous multiresolution mesh modeling techniques suitable for gaze-contingent viewing have recently been developed (Zorin & Schröder, 2000). Techniques range from multiresolution representation of arbitrary meshes to the management of LOD through peripheral degradation within an HMD where gaze position is assumed to coincide with head direction (Lindstrom et al, 1996; MacCracken & Joy, 1996; Hoppe, 1997; Zorin, Schröder, & Sweldens, 1997; Schmalstieg & Schaufler, 1997). Although some of these authors address view and gaze dependent object representation, few results concerning display speedup are as yet available showing successful adaptation of these techniques within a true gaze-contingent system, i.e., one where an eye tracker is employed. Due to the advancements of multiresolution modeling techniques and to the increased affordability of eye trackers, it is now becoming feasible to extend the LOD approach to gaze-contingent displays, where models are rendered *nonisotropically*.

An early example of a nonisotropic model-based gaze-contingent system, where gaze direction is directly applied to the rendering algorithm, was presented by Levoy and Whitaker (1990). The authors' spatially adaptive near real-time ray tracer for volume data displayed an eye-slaved ROI by modulating both the number of rays cast per unit area on the image plane and the number of samples drawn per unit length along each ray as a function of local retinal acuity. The ray-traced image was sampled by a nonisotropic convolution filter to generate a 12° foveal ROI within a 20° mid-resolution transitional region. Based on preliminary estimates, the authors suggested a reduction in image generation time by a factor of up to 5. An NAC Eye Mark eye tracker was used to determine the user's POR while viewing a conventional 19″ TV monitor. A chin rest and immobilization strap were used to eliminate the need for head tracking.

For environments containing significant topological detail, such as virtual terrains, rendering with multiple levels of detail, where the level is based on user position and gaze direction, is essential to provide an acceptable combination of surface detail and frame rate (Danforth, Duchowski, Geist, & McAliley, 2000). Recent work in this area has been extensive. Particularly impressive is Hoppe's (1998) view-dependent progressive mesh framework, where spatial continuity is maintained through structure design, and temporal continuity is maintained by *geomorphs*.

Danforth et al (2000) used an eye tracker as an indicator of gaze in a gaze-contingent multiresolution terrain navigation environment. A surface, represented as a quadrilateral mesh, was divided into fixed-size (number of vertices) sub-blocks, allowing rendering for variable LOD on a per-sub-block basis. From a fully-detailed surface, lower levels of resolution are constructed by removing half of the vertices in each direction and assigning new vertex values. The new values are averages of the higher resolution values. Resolution level was chosen per sub-block, based on viewer distance. The resolution level was not discrete; it was interpolated between the pre-computed discrete levels to avoid "popping" effects. The terrain, prior to gaze-contingent alteration, is shown in Figure 14.6. Rocks in the terrain are rendered by bill-boarding, i.e., images of rocks from the Pathfinder mission to Mars were rendered onto 2D transparent planes that rotate to maintain an orientation orthogonal to the viewer. Two views of the gaze-contingent environment (shown rendered and in wireframe) are seen in Figures 14.7 and 14.8. To exaggerate the gaze-contingent effect, in this environment, fractal mountains appear and disappear from view, based on direction of gaze. Notice also, in Figures 14.7(a) and (b), the increased resolution (number of quads) below the gaze vector. The images in the figure are snapshots of the scene images generated by the eye tracker, i.e., what is seen by the operator–the point of regard cross-hair, coordinates, and video frame timecode are not seen by the viewer immersed in the environment.

More recent work on gaze-contingent LOD modeling has been carried out by Luebke and Erikson (1997). The authors present a view-dependent LOD technique suitable for gaze-contingent rendering. While simplification of individual geometric objects is discussed in their work, it appears the strategy is ultimately directed toward solving the interactive "walkthrough" problem (Funkhouser & Séquin, 1993). In this application, the view-dependent LOD technique seems more suitable to the (possibly) gaze-contingent rendering of an entire scene or environment. Recently, the authors have developed a gaze-

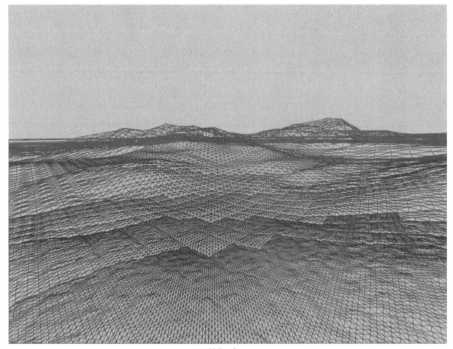

(a) Wireframe.

(b) Rendered.

Fig. 14.6. Fractal terrain for gaze-contingent virtual environment. Courtesy of Bob Danforth.

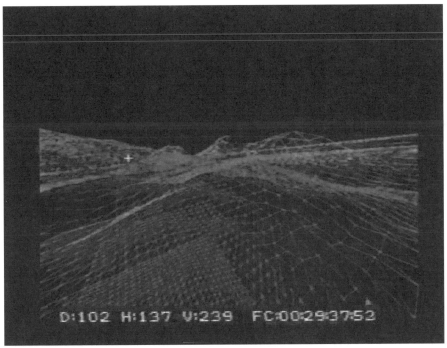

(a) Looking left.

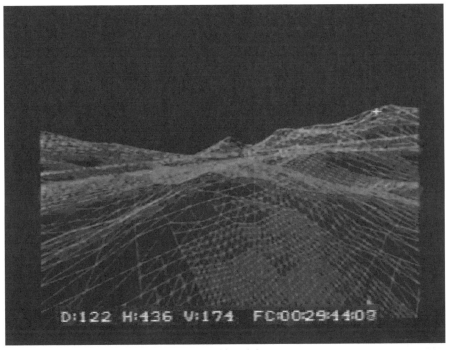

(b) Looking right.

Fig. 14.7. Fractal terrain: gaze-contingent rendering (wireframe).

(a) Looking left.

(b) Looking right.

Fig. 14.8. Fractal terrain: gaze-contingent rendering.

directed LOD technique to facilitate the gaze-contingent display of geometric objects (Luebke, Hallen, Newfield, & Watson, 2000). To test their rendering approach the authors employed a table-mounted monocular eye tracker to measure the viewer's real-time location of gaze over a desktop display. While this work shows the feasibility of employing an eye tracker, the implementation framework used by the authors lacked a head tracker and required a chin rest to ensure tracker accuracy.

A new object-based LOD method has been developed by Murphy and Duchowski (2001). The technique is similar to Luebke and Erikson's and to Ohshima et al's, where objects are modeled for gaze-contingent viewing. Unlike the approach of Ohshima et al, resolution degradation is applied nonisotropically, i.e., objects are not necessarily degraded uniformly. The spatial degradation function for LOD selection differs significantly from the area-based criteria originally proposed by Clarke. Instead of evaluating the screen coverage of the projected object, the degradation function is based on the evaluation of visual angle in world coordinates. System performance measurements are reported, obtained from experiments using a binocular eye tracker built into an HMD. Tracking software obtains helmet position and orientation in real-time and calculates the direction of the user's gaze. The geometric modeling technique developed for the purpose of gaze-contingent rendering includes an integrated approach to tiling, mapping, and remeshing of closed surfaces. A three-dimensional spatial degradation function was obtained from human subject experiments in an attempt to imperceptibly display spatially degraded geometric objects. System performance measurements indicate an approximate overall 10-fold average frame rate improvement during gaze-contingent viewing. Two frames during gaze-contingent viewing of one geometric model is shown in Figure 14.9.

Another interesting approach to gaze-contingent modeling for real-time graphics rendering was taken by O'Sullivan and Dingliana (2001) and O'Sullivan, Dingliana, and Howlett (2002). Instead of degrading the resolution of peripherally located geometric objects, O'Sullivan and Dingliana (2001) consider a degradable collision handling mechanism to effectively limit object collision resolution outside a foveal Region Of Interest (ROI). When the viewer is looking directly at a collision, it is given higher (computational) priority than collisions occurring in the periphery. Object collisions given higher priority are allocated more processing time so that the contact model and resulting response is more believable. The calculated collision response accuracy (i.e., the precision of the calculated collision point between two rigid objects) is directly

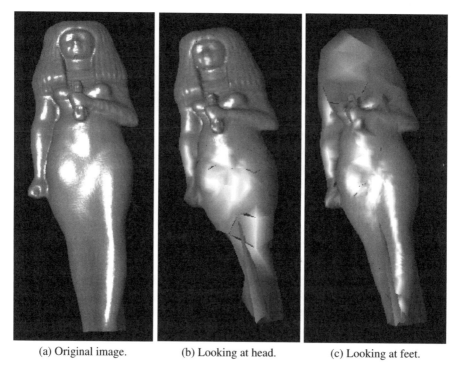

(a) Original image.   (b) Looking at head.   (c) Looking at feet.

Fig. 14.9. Gaze-contingent viewing of *isis* model. Courtesy of Hunter Murphy.

dependent on the level of collision detection detail. A lower level of detail results in less physically accurate physics and objects not touching when they bounce, leaving a potentially perceivable gap.

O'Sullivan and Dingliana (2001) describe psychophysical experiments to measure the perceptibility of coarsely simulated collisions in the periphery. The authors tested effects of factors such as eccentricity, separation, presence and number of similar and dissimilar distractors, causality, and physics on participants' perception of collisions. To test viewers' sensitivity to collision resolution (size of gap between colliding objects) at eccentricity, O'Sullivan and Dingliana tested whether the ability of viewers to detect anomalous collisions, in this case colliding objects that do not touch each other but leave a gap, decreases with increasing eccentricity of the collision point. Collisions were presented at 5 eccentricities of $1.4°$, $2.9°$, $4.3°$, $5.7°$, and $7.2°$ visual angle. The authors note a significant fall-off in detection accuracy with eccentricity (at about $4°$ visual angle).

Based on their previous psychophysical findings, O'Sullivan et al (2002) developed a gaze-contingent collision handling system. Two variants of the gaze-contingent system were compared, each containing a high-priority ROI wherein collisions were processed at higher resolution than outside the ROI. In the tracked case, the high-priority ROI was synchronized to the viewer's tracked gaze position with an SMI EyeLink eye tracker, while in the random case, the ROI position was determined randomly every 5 frames. An example of the graphics simulation with highlighted ROI and subject wearing an eye tracker are shown in Figure 14.10. O'Sullivan et al report an overall improvement in the perception of the tracked simulation.

14.3 Summary and Further Reading

Several classes of eye tracking applications were presented falling generally within the domain of Computer Science and Human-Computer Interaction. HCI-related studies have included some of the first well-known adaptations of eye trackers to computer-based systems. This trend will probably continue for some time. Interactive eye tracking systems including those featured in Computer-Supported Collaborative Work (CSCW) will most likely continue to be explored for directly interactive support (e.g., gaze pointing) as well as for indirect interaction (e.g., gaze-assisted pointing). However, the recent interest in diagnostic uses of eye trackers, especially in usability studies, is also expected to flourish.

For an excellent review of Gaze-Contingent Multi-Resolution Displays, or GCMRDs, see Reingold, Loschky, McConkie, and Stampe (2002) as well as Parkhurst and Niebur (2002). The remainder of gaze-contingent work described in this chapter mostly emanates from research in Computer Graphics. Gaze-contingent displays described here are usually interactive, real-time examples of adoption of eye trackers into graphical displays. See O'Sullivan et al (2002) for a review of eye tracking work in interactive graphics. Here too, eye trackers are becoming noticed for their diagnostic contributions. The recent ACM SIGGRAPH Campfire on Perceptually Adaptive Graphics (McNamara & O'Sullivan, 2001) suggests that general perceptual issues are becoming particularly important in computer graphics research. Traditional graphics algorithms (e.g., ray-tracing, radiosity) are beginning to include perceptually-based improvements to further speed up processing time. Eye trackers will certainly be able to contribute in a diagnostic capacity to improve the fidelity of perceptually-enhanced graphical imagery.

(a) Important collisions (e.g., those close to the viewer's fixation position) are processed first.

(b) An eye-tracker is used to determine the viewer's point of fixation.

Fig. 14.10. Gaze-contingent collision modeling. Images courtesy of C. O'Sullivan and J.Dingliana.

For further reading, the two best sources of current research in HCI and Computer Graphics are the annual ACM SIGCHI and ACM SIGGRAPH conferences and the related ACM journal articles in the Transactions on Computer-Human Interaction and the Transactions on Graphics. Work specifically related to Virtual Reality can be found in the ACM VRST and IEEE VR conferences as well as in the journal Presence.

15. Conclusion

Eye tracking has sometimes been referred to as a technology in search of an application. Through the presentation of examples in this text, which is by no means exhaustive, it appears that many interesting applications have now been found. There is a good deal of opportunity for interesting, meaningful research.

It has also been said that eye trackers are difficult to set up, use, and that they are unwieldy and expensive. There may be some truth in this observation. Through ongoing improvements to the technology, eye trackers are becoming much more easier to use. Video-based eye trackers in particular are becoming relatively inexpensive, quite accurate, and fairly easy to use. The current "holy grail" of eye tracking research is a calibration-free eye tracker. Several research centers are pursuing this goal. With the use of multiple video cameras and computer vision techniques, this goal may be achieved very soon.

Set up and design of supporting computer programs and methodologies for eye tracking studies may still require some expertise and thus may pose some challenges. Hopefully this book has in some way been useful in addressing these challenges. The contents of the text primarily deal with the computational "back-end" methodological issues, e.g., program design for graphical and Virtual Reality displays and subsequent eye movement collection and analysis. Often a complete eye tracking study may require collaboration with members from several traditionally distinct disciplines. What has seemed to work in the classroom may also work in the field, the lab, or on campus, and that is the formation of interdisciplinary teams. The assembly of members from say Psychology, Marketing, Industrial Engineering, and Computer Science may be the most effective strategy for conducting effective eye tracking research.

References

Allopenna, P. D., Magnuson, J. S., & Tanenhaus, M. K. (1998). Tracking the Time Course of Spoken Word Recognition Using Eye Movements: Evidence for Continuous Mapping Models. *Journal of Memory and Language*, *38*(4), 419-439.

Ancman, C. E. G. (1991). Peripherally Located CRTs: Color Perception Limitations. In *National Aerospace and Electronics Conference (NAECON)* (p. 960-965). Dayton, OH.

Anders, G. (2001). Pilot's Attention Allocation During Approach and Landing Eye- and Head-Tracking Research in an A330 Full Flight Simulator. In *International Symposium on Aviation Psychology (ISAP)*. Columbus, OH. (URL: <http://www.geerdanders.de/literatur/2001_ohio.html>. Last accessed 12/30/01.)

Anderson, J. R. (1993). *Rules of the Mind*. Hillsdale, NJ: Lawrence Erlbaum Associates.

Anderson, J. R. (2002). *Personal Communique*.

Anliker, J. (1976). Eye Movements: On-Line Measurement, Analysis, and Control. In R. A. Monty & J. W. Senders (Eds.), *Eye Movements and Psychological Processes* (p. 185-202). Hillsdale, NJ: Lawrence Erlbaum Associates.

Arbib, M. A. (Ed.). (1995). *The Handbook of Brain Theory and Neural Networks*. Cambridge, MA: The MIT Press.

Asaad, W. F., Rainer, G., & Miller, E. K. (2000). Task-Specific Neural Activity in the Primate Prefrontal Cortex. *Neurophysiology*, *84*, 451-459.

Bahill, A. T., Clark, M., & Stark, L. (1975). The Main Sequence, A Tool for Studying Human Eye Movements. *Mathematical Biosciences*, *24*(3/4), 191-204.

Ballard, D. H., Hayhoe, M. M., & Pelz, J. B. (1995). Memory Representations in Natural Tasks. *Journal of Cognitive Neuroscience*, *7*(1), 66-80.

Bass, M. (Ed.). (1995). *Handbook of Optics: Fundamentals, Techniques, & Design (2nd edition)*. New York, NY: McGraw-Hill, sponsored by the Optical Society of America.

Becker, W. (1989). Metrics. In R. H. Wurtz & M. E. Goldberg (Eds.), *The Neurobiology of Saccadic Eye Movements* (p. 13-68). New York, NY: Elsevier Science Publishers BV (Biomedical Division).

Bertera, J. H., & Rayner, K. (2000). Eye Movements and the Span of the Effective Stimulus in Visual Search. *Perception & Psychophysics, 62*(3), 576-585.

Bloom, F. E., & Lazerson, A. (1988). *Brain, Mind, and Behavior* (2nd ed.). New York, NY: W. H. Freeman and Company.

Boff, K. R., & Lincoln, J. E. (Eds.). (1988). *Engineering Data Compendium: Human Perception and Performance.* Wright-Patterson AFB, OH: USAF Harry G. Armstrong Aerospace Medical Research Laboratory (AAMRL).

Brinkmann, R. (1999). *The Art and Science of Digital Compositing.* New York, NY: Morgan Kaufmann.

Broadbent, D. E. (1958). *Perception and Communication.* Oxford: Pergamon Press.

Buswell, G. T. (1935). *How People Look At Pictures.* Chicago, IL: University of Chicago Press.

Byrne, M. D., Anderson, J. R., Douglass, S., & Matessa, M. (1999). Eye Tracking the Visual Search of Click-Down Menus. In *Human Factors in Computing Systems: CHI '99 Conference Proceedings* (p. 402-409). ACM Press.

Carpenter, R. H. S. (1977). *Movements of the Eyes.* London: Pion Limited.

Chapman, P. R., & Underwood, G. (1998). Visual Search of Dynamic Scenes: Event Types and the Role of Experience in Viewing Driving Situations. In G. Underwood (Ed.), *Eye Guidance in Reading and Scene Perception* (p. 369-394). Oxford, England: Elsevier.

Clark, M. R., & Stark, L. (1975). Time Optimal Behavior of Human Saccadic Eye Movement. *IEEE Transactions on Automatic Control, 20*, 345-348.

Clarke, J. H. (1976). Hierarchical Geometric Models for Visible Surface Algorithms. *Communications of the ACM, 19*(10), 547-554.

Cooper, R. M. (1974). The Control of Eye Fixation by the Meaning of Spoken Language: A New Methodology for the Real-Time Investigation of Speech Perception, Memory, and Language Processing. *Cognitive Psychology, 6*, 84-107.

Cornsweet, T. N. (1970). *Visual Perception.* New York, NY: Academic Press.

Crane, H. D. (1994). The Purkinje Image Eyetracker, Image Stabilization, and Related Forms of Stimulus Manipulation. In D. H. Kelly (Ed.), *Visual Science and Engineering: Models and Applications* (p. 13-89). New York, NY: Marcel Dekker, Inc.

Crane, H. D., & Steele, C. M. (1985). Generation-V Dual-Purkinje-Image Eyetracker. *Applied Optics, 24*, 527-537.

Crichton, M. W. (1981). *Looker [Motion Picture]*. Warner Brothers (USA).

Crundall, D. E., Underwood, G., & Chapman, P. R. (1998). How Much Do Novice Drivers See? The Effects of Demand on Visual Search Strategies in Novice and Experienced Drivers. In G. Underwood (Ed.), *Eye Guidance in Reading and Scene Perception* (p. 395-418). Oxford, England: Elsevier.

Danforth, R., Duchowski, A., Geist, R., & McAliley, E. (2000). A Platform for Gaze-Contingent Virtual Environments. In *Smart Graphics (Papers from the 2000 AAAI Spring Symposium, Technical Report SS-00-04)* (p. 66-70). Menlo Park, CA.

Davson, H. (1980). *Physiology of the Eye* (4th ed.). New York, NY: Academic Press.

Deutsch, J. A., & Deutsch, D. (1963). Attention: Some Theoretical Considerations. *Psychological Review, 70*(1), 80-90.

De Valois, R. L., & De Valois, K. K. (1988). *Spatial Vision*. New York, NY: Oxford University Press.

Dishart, D. C., & Land, M. F. (1998). The Development of the Eye Movement Strategies of Learner Drivers. In G. Underwood (Ed.), *Eye Guidance in Reading and Scene Perception* (p. 419-430). Oxford, England: Elsevier.

Dodge, R. (1907). An Experimental Study of Visual Fixation. *Psychological Monograph, 8*(4).

Doll, T. J. (1993). Preattentive Processing in Visual Search. In *Proceedings of the Human Factors and Ergonomics Society, 37th Annual Meeting* (p. 1291-1249). Santa Monica, CA.

Doll, T. J., Whorter, S. W., & Schmieder, D. E. (1993). Simulation of Human Visual Search in Cluttered Backgrounds. In *Proceedings of the Human Factors and Ergonomics Society, 37th Annual Meeting* (p. 1310-1314). Santa Monica, CA.

Dowling, J. E., & Boycott, B. B. (1966). Organization of the Primate Retina: Electron Microscopy. *Proceedings of the Royal Society (London), 166*, 80-111. (Series B)

Doyal, J. A. (1991). *Spatial Summation for Color in the Peripheral Retina*. Unpublished master's thesis, Wright State University.

Drury, C. G., Gramopadhye, A. K., & Sharit, J. (1997). Feedback Strategies for Visual Inspection in Airframe Structural Inspection. *International Journal of Industrial Ergonomics, 19*, 333-344.

Duchowski, A., Medlin, E., Cournia, N., Gramopadhye, A., Melloy, B., & Nair, S. (2002). 3D Eye Movement Analysis for VR Visual Inspection Training. In *Eye Tracking Research & Applications (ETRA)* (pp. 103-110,155). New Orleans, LA: ACM.

Duchowski, A., Medlin, E., Gramopadhye, A., Melloy, B., & Nair, S. (2001). Binocular Eye Tracking in VR for Visual Inspection Training. In *Virtual Reality Software & Technology (VRST)*. Banff, AB, Canada. (To appear.)

Duchowski, A. T. (2000). Acuity-Matching Resolution Degradation Through Wavelet Coefficient Scaling. *IEEE Transactions on Image Processing*, *9*(8), 1437-1440.

d'Ydewalle, G., Desmet, G., & Van Rensbergen, J. (1998). Film Perception: The Processing of Film Cuts. In G. Underwood (Ed.), *Eye Guidance in Reading and Scene Perception* (p. 357-368). Oxford, England: Elsevier.

Findlay, J. M. (1992). Programming of Stimulus-Elicited Saccadic Eye Movements. In K. Rayner (Ed.), *Eye Movements and Visual Cognition: Scene Perception and Reading* (p. 8-30). New York, NY: Springer-Verlag. (Springer Series in Neuropsychology)

Findlay, J. M. (1997). Saccade Terget Selection During Visual Search. *Vision Research*, *37*(5), 617-631.

Findlay, J. M., & Gilchrist, I. D. (1998). Eye Guidance and Visual Search. In G. Underwood (Ed.), *Eye Guidance in Reading and Scene Perception* (p. 295-312). Oxford, England: Elsevier.

Finke, R. A. (1989). *Principles of Mental Imagery*. Cambridge, MA: The MIT Press.

Fisher, D. F., Monty, R. A., & Senders, J. W. (Eds.). (1981). *Eye Movements: Cognition and Visual Perception*. Hillsdale, NJ: Lawrence Erlbaum Associates.

Fuchs, A. F., Kaneko, C. R. S., & Scudder, C. A. (1985). Brainstem Control of Saccadic Eye Movements. *Annual Review of Neuroscience*, *8*, 307-337.

Funkhouser, T. A., & Séquin, C. H. (1993). Adaptive Display Algorithm for Interactive Frame Rates During Visualization of Complex Virtual Environments. In *Computer Graphics (SIGGRAPH '93)*. New York, NY.

Gamlin, P., & Twieg, D. (1997). *A Combined Visual Display and Eye Tracking System for High-Field FMRI Studies*. (NSF Interactive Systems Grantees Workshop (ISGW'97) (report abstract available at URL: <http://www.cse.ogi.edu/CSLU/isgw97/reports/gamlin.html>, last referenced March 1998))

Gazzaniga, M. S. (Ed.). (2000). *The New Neurosceinces*. Cambridge, MA: The MIT Press.

Gibson, J. J. (1941). A Critical Review of the Concept of Set in Contemporary Experimental Psychology. *Psychological Bulletin*, *38*(9), 781-817.

Glassner, A. S. (Ed.). (1989). *An Introduction to Ray Tracing.* San Diego, CA: Academic Press.

Goldberg, J., Stimson, M., Lewnstein, M., Scott, N., & Wichansky, A. (2002). Eye Tracking in Web Search Tasks: Design Implications. In *Eye Tracking Research & Applications (ETRA) Symposium.* New Orleans, LA.

Goldberg, J. H., & Kotval, X. P. (1999). Computer Interface Evaluation Using Eye Movements: Methods and Constructs. *International Journal of Industrial Ergonomics, 24,* 631-645.

Graeber, D. A., & Andre, A. D. (1999). Assessing Visual Attention of Pilots While Using Electronic Moving Maps for Taxiing. In R. S. Jensen, B. Cox, J. D. Callister, & R. Lavis (Eds.), *International Symposium on Aviation Psychology (ISAP)* (p. 791-796). Columbus, OH.

Gramopadhye, A., Bhagwat, S., Kimbler, D., & Greenstein, J. (1998). The Use of Advanced Technology for Visual Inspection Training. *Applied Ergonomics, 29*(5), 361-375.

Greene, H. H., & Rayner, K. (2001). Eye Movements and Familiarity Effects in Visual Search. *Vision Research, 41*(27), 3763-3773.

Gregory, R. L. (1990). *Eye and Brain: The Psychology of Seeing.* Princeton, NJ: Princeton University Press.

Grzywacz, N. M., & Norcia, A. M. (1995). Directional Selectivity in the Cortex. In M. A. Arbib (Ed.), *The Handbook of Brain Theory and Neural Networks* (p. 309-311). Cambridge, MA: The MIT Press.

Grzywacz, N. M., Sernagor, E., & Amthor, F. R. (1995). Directional Selectivity in the Retina. In M. A. Arbib (Ed.), *The Handbook of Brain Theory and Neural Networks* (p. 312-314). Cambridge, MA: The MIT Press.

Haber, R. N., & Hershenson, M. (1973). *The Psychology of Visual Perception.* New York, NY: Holt, Rinehart, and Winston, Inc.

Hain, T. C. (1999). *Saccade (Calibration) Tests.* (Online Manual, URL: <http://www.tchain.com/otoneurology/practice/saccade.htm> (last accessed October 2001))

Hayhoe, M. M., Ballard, D. H., Triesch, J., Shinoda, H., Rogriguez, P. A., & Sullivan, B. (2002). Vision in Natural and Virtual Environments (Invited Paper). In *Eye Tracking Research & Applications (ETRA) Symposium.* New Orleans, LA.

He, P., & Kowler, E. (1989). The Role of Location Probability in the Programming of Saccades: Implications for "Center-of-Gravity" Tendencies. *Vision Research, 29*(9), 1165-1181.

Heijden, A. H. C. Van der. (1992). *Selective Attention in Vision.* London: Routledge.

Hendee, W. R., & Wells, P. N. T. (1997). *The Perception of Visual Information* (2nd ed.). New York, NY: Springer-Verlag, Inc.

Henderson, J. M. (1992). Object Identification in Context: The Visual Processing of Natural Scenes. *Canadian Journal of Psychology, 46*(3), 319-341.

Henderson, J. M., & Hollingworth, A. (1998). Eye Movements During Scene Viewing: An Overview. In G. Underwood (Ed.), *Eye Guidance in Reading and Scene Perception* (p. 269-294). Oxford, England: Elsevier.

Ho, G., Scialfa, C. T., Caird, J. K., & Graw, T. (2001). Visual Search for Traffic Signs: The Effects of Clutter, Luminance, and Aging. *Human Factors, 43*(3), 194-207.

Hoppe, H. (1997). View-Dependent Refinement of Progressive Meshes. In *Computer Graphics (SIGGRAPH '97)*. New York, NY.

Hoppe, H. (1998). Smooth View-Dependent Level-Of-Detail Control and its Application to Terrain Rendering. In *Proceedings of IEEE Visualization* (p. 35-42). Raleigh, NC.

Hornof, A. J., & Kieras, D. E. (1997). Cognitive Modeling Reveals Menu Search is Both Random and Systematic. In *Human Factors in Computing Systems: CHI '97 Conference Proceedings* (p. 107-144). Atlanta, GA.

Hubel, D. H. (1988). *Eye, Brain, and Vision*. New York, NY: Scientific American Library.

Hughes, H. C., Nozawa, G., & Kitterle, F. (1996). Global Precedence, Spatial Frequency Channels, and the Statistics of Natural Images. *Journal of Cognitive Neuroscience, 8*(3), 197-230.

Hutchinson, T. E. (1993). Eye-gaze computer interfaces. *IEEE Computer, 26*(7), 65,67.

Irwin, D. E. (1992). Visual Memory Within and Across Fixations. In K. Rayner (Ed.), *Eye movements and visual cognition: Scene perception and reading* (p. 146-165). New York, NY: Springer-Verlag. (Springer Series in Neuropsychology)

Itti, L., Koch, C., & Niebur, E. (1998). A Model of Saliency-Based Visual Attention for Rapid Scene Analysis. *IEEE Transactions on Pattern Analysis and Machine Intelligence (PAMI), 20*(11), 1254-1259.

Jacob, R. J. (1990). What You Look at is What You Get: Eye Movement-Based Interaction Techniques. In *Human Factors in Computing Systems: CHI '90 Conference Proceedings* (p. 11-18). ACM Press.

James, W. (1981). *The Principles of Psychology* (Vol. I). Cambridge, MA: Harvard University Press. ((See: James, William, *The Principles of Psychology*, H. Holt and Co., New York, NY, 1890.))

Kanizsa, G. (1976). Subjective Contours. *Scientific American, 234*(4), 48-52,138.

Kaplan, E. (1991). The Receptive Field Structure of Retinal Ganglion Cells in Cat and Monkey. In A. G. Leventhal & J. R. Cronly-Dillon (Eds.), *The Neural Basis of Visual Function* (p. 10-40). Boca Raton, FL: CRC Press. (Vision and Visual Dysfunction Series, vol. 4)

Karn, K. S., Ellis, S., & Juliano, C. (1999). The Hunt for Usability: Tracking Eye Movements. In *Human Factors in Computing Systems: CHI '99 Conference Proceedings (Workshops)* (p. 173). ACM Press.

Kennedy, A. (1992). The Spatial Coding Hypothesis. In K. Rayner (Ed.), *Eye Movements and Visual Cognition: Scene Perception and Reading* (p. 379-396). New York, NY: Springer-Verlag. (Springer Series in Neuropsychology)

Kieras, D., & Meyer, D. E. (1995). *An Overview of the EPIC Architecture for Cognition and Performance with Appliaction to Human-Computer Interaction* (EPIC Tech. Rep. No. 5 No. TR-95/ONR-EPIC-5). Ann Arbor, University of Michigan, Electrical Engineering and Computer Science Department.

Knox, P. C. (2001). *The Parameters of Eye Movement.* (Lecture Notes, URL: <http://www.liv.ac.uk/ pcknox/teaching/Eymovs/params.htm> (last accessed October 2001))

Kocian, D. (1987). Visual World Subsystem. In *Super Cockpit Industry Days: Super Cockpit/Virtual Crew Systems.* Air Force Museum, Wright-Patterson AFB, OH.

Koenderink, J. J., van Doorn, A. J., & van de Grind, W. A. (1985). Spatial and Temporal Parameters of Motion Detection in the Peripheral Visual Field. *J. Opt. Soc. Am., 2*(2), 252-259.

Kortum, P., & Geisler, W. S. (1996). Implementation of a foveated image coding system for bandwidth reduction of video images. In *Human vision and electronic imaging* (p. 350-360). Bellingham, WA.

Kosslyn, S. M. (1994). *Image and Brain.* Cambridge, MA: The MIT Press.

Kroll, J. F. (1992). Making a Scene: The Debate about Context Effects for Scenes and Sentences. In K. Rayner (Ed.), *Eye Movements and Visual Cognition: Scene Perception and Reading.* Springer-Verlag. (Springer Series in Neuropsychology)

Land, M., Mennie, N., & Rusted, J. (1999). The Roles of Vision and Eye Movements in the Control of Activities of Daily Living. *Perception, 28*(11), 1307-1432.

Land, M. F., & Hayhoe, M. (2001). In What Ways Do Eye Movements Contribute to Everyday Activities. *Vision Research, 41*(25-26), 3559-3565.

((Special Issue on *Eye Movements and Vision in the Natual World*, with most contributions to the volume originally presented at the 'Eye Movements and Vision in the Natural World' symposium held at the Royal Netherlands Academy of Sciences, Amsterdam, September 2000))

Laurutis, V. P., & Robinson, D. A. (1986). The Vestibulo-ocular Reflex During Human Saccadic Eye Movements. *Journal of Physiology, 373*, 209-233.

Leigh, R. J., & Zee, D. S. (1991). *The Neurology of Eye Movements* (2nd ed.). Philadelphia, PA: F. A. Davis Company.

Levoy, M., & Whitaker, R. (1990). Gaze-Directed Volume Rendering. In *Computer Graphics (SIGGRAPH '90)* (p. 217-223). New York, NY.

Lindstrom, P., Koller, D., Ribarsky, W., Hodges, L. F., Faust, N., & Turner, G. A. (1996). Real-Time, Continuous Level of Detail Rendering of Height Fields. In *Computer Graphics (SIGGRAPH '96)* (p. 109-118). New York, NY.

Liu, A. (1998). What the Driver's Eye Tells the Car's Brain. In G. Underwood (Ed.), *Eye Guidance in Reading and Scene Perception* (p. 431-452). Oxford, England: Elsevier.

Livingstone, M., & Hubel, D. (1988). Segregation of Form, Color, Movement, and Depth: Anatomy, Physiology, and Perception. *Science, 240*, 740-749.

Loftus, G. R. (1981). Tachistoscopic Simulations of Eye Fixations on Pictures. *Journal of Experimental Psychology: Human Learning and Memory, 7*(5), 369-376.

Logothetis, N. K., & Leopold, D. A. (1995). *On the Physiology of Bistable Percepts* (AI Memo No. 1553). Artificial Intelligence Laboratory, Massachusetts Institute of Technology.

Lohse, G. L. (1997). Consumer Eye Movement Patterns on Yellow Pages Advertising. *Journal of Advertising, 26*(1), 61-73.

Longridge, T., Thomas, M., Fernie, A., Williams, T., & Wetzel, P. (1989). Design of an Eye Slaved Area of Interest System for the Simulator Complexity Testbed. In T. Longridge (Ed.), *Area of Interest/Field-Of-View Research Using ASPT (Interservice/Industry Training Systems Conference)* (p. 275-283). Brooks Air Force Base, TX: Air Force Human Resources Laboratory, Air Force Systems Command.

Loschky, L. C., & McConkie, G. W. (2000). User Performance With Gaze Contingent Multiresolutional Displays. In *Eye Tracking Research & Applications Symposium* (p. 97-103). Palm Beach Gardens, FL.

Luebke, D., & Erikson, C. (1997). View-Dependent Simplification Of Arbitrary Polygonal Environments. In *Computer Graphics (SIGGRAPH '97).* New York, NY.

Luebke, D., Hallen, B., Newfield, D., & Watson, B. (2000). *Perceptually Driven Simplification Using Gaze-Directed Rendering* (Tech. Rep. No. CS-2000-04). University of Virginia.

Lund, J. S., Wu, Q., & Levitt, J. B. (1995). Visual Cortex Cell Types and Connections. In M. A. Arbib (Ed.), *The Handbook of Brain Theory and Neural Networks* (p. 1016-1021). Cambridge, MA: The MIT Press.

MacCracken, R., & Joy, K. (1996). Free-From Deformations With Lattices of Arbitrary Topology. In *Computer Graphics (SIGGRAPH '96)* (p. 181-188). New York, NY.

Majaranta, P., & Raiha, K.-J. (2002). Twenty Years of Eye Typing: Systems and Design Issues. In *Eye Tracking Research & Applications (ETRA) Symposium.* New Orleans, LA.

McColgin, F. H. (1960). Movement Thresholds in Peripheral Vision. *Journal of the Optical Society of Amercia*, 50(8), 774-779.

McConkie, G. W., & Rayner, K. (1975). The Span of the Effective Stimulus During a Fixation in Reading. *Perception & Psychophysics*, 17, 578-586.

McCormick, B. H., Batte, D. A., & Duchowski, A. T. (1996). *A Virtual Environment: Exploring the Brain Forest.* (Presentation at the CENAC Conference, October 1996, Mexico City, Mexico (preprint available from principal author, Department of Computer Science, Texas A&M University, College Station, TX))

McNamara, A., & O'Sullivan, C. (Eds.). (2001). *Perceptually Adaptive Graphics.* Snowbird, UT: ACM SIGGRAPH/EuroGraphics. (URL: <http://isg.cs.tcd.ie/campfire/>, last referenced May 2002)

Megaw, E. D., & Richardson, J. (1979). Eye Movements and Industrial Inspection. *Applied Ergonomics*, 10, 145-154.

Menache, A. (2000). *Understanding Motion Capture for Computer Animation and Video Games.* San Diego, CA: Morgan Kaufmann, (Academic Press).

Molnar, F. (1981). About the Role of Visual Exploration in Aesthetics. In H. Day (Ed.), *Advances in Intrinsic Motivation and Aesthetics.* New York, NY: Plenum Press.

Monty, R. A., & Senders, J. W. (Eds.). (1976). *Eye Movements and Psychological Processes.* Hillsdale, NJ: Lawrence Erlbaum Associates.

Murphy, H., & Duchowski, A. T. (2001). Gaze-Contingent Level Of Detail. In *Eurographics (short presentations).* Manchester, UK.

Necker, L. A. (1832). Observations on Some Remarkable Optical Phaenomena Seen in Switzerland, and on an Optical Phaenomenon Which Occurs on

Viewing a Figure or a Crystal or Geometrical Solid. *Phiolosophical Magazine and Journal of Science, 1*, 329-337. (Third Series.)

Nguyen, E., Labit, C., & Odobez, J.-M. (1994). A *ROI* Approach for Hybrid Image Sequence Coding. In *International Conference on Image Processing (ICIP)'94* (p. 245-249). Austin, TX.

Nielsen, J. (1993). The Next Generation GUIs: Noncommand User Interfaces. *Communications of the ACM, 36*(4), 83-99.

Ninio, J., & Stevens, K. A. (2000). Variations on the Hermann Grid: An Extinction Illusion. *Perception, 29*, 1209-1217.

Noton, D., & Stark, L. (1971a). Eye Movements and Visual Perception. *Scientific American, 224*, 34-43.

Noton, D., & Stark, L. (1971b). Scanpaths in Saccadic Eye Movements While Viewing and Recognizing Patterns. *Vision Research, 11*, 929-942.

Ohshima, T., Yamamoto, H., & Tamura, H. (1996). Gaze-Directed Adaptive Rendering for Interacting with Virtual Space. In *Proceedings of VRAIS'96* (p. 103-110). Santa Clara, CA.

O'Regan, K. J. (1992). Optimal Viewing Position in Words and the Strategy-Tactics Theory of Eye Movements in Reading. In K. Rayner (Ed.), *Eye Movements and Visual Cognition: Scene Perception and Reading* (p. 333-355). New York, NY: Springer-Verlag. (Springer Series in Neuropsychology)

Osberger, W., & Maeder, A. J. (1998). Automatic Identification of Perceptually Important Regions in an Image. In *International Conference on Pattern Recognition.* Brisbane, Australia.

O'Sullivan, C., & Dingliana, J. (2001). Collisions and Perception. *ACM Transactions on Graphics, 20*(3).

O'Sullivan, C., Dingliana, J., & Howlett, S. (2002). Gaze-Contingent Algorithms for Interactive Graphics. In J. Hyöna, R. Radach, & H. Deubel (Eds.), *The Mind's Eyes: Cognitive and Applied Aspects of Eye Movement Research.* Oxford, England: Elsevier Science.

Ottati, W. L., Hickox, J. C., & Richter, J. (1999). Eye Scan Patterns of Experienced and Novice Pilots During Visual Flight Rules (VFR) Navigation. In *Proceedings of the Human Factors and Ergonomics Society, 43rd Annual Meeting.* Houston, TX.

Özyurt, J., DeSouza, P., West, P., Rutschmann, R., & Greenlee, M. W. (2001). Comparison of Cortical Activity and Oculomotor Performance in teh Gap and Step Paradigms. In *European Conference on Visual Perception (ECVP).* Kusadasi, Turkey.

Palmer, S. (1999). *Vision Science.* Cambridge, MA: MIT Press.

Papathomas, T. V., Chubb, C., Gorea, A., & Kowler, E. (Eds.). (1995). *Early Vision and Beyond.* Cambridge, MA: The MIT Press.

Parkhurst, D., Culurciello, E., & Niebur, E. (2000). Evaluating Variable Resolution Displays with Visual Search: Task Performance and Eye Movements. In *Eye Tracking Research & Applications (ETRA)* (p. 105-109). Palm Beach Gardens, FL.

Parkhurst, D. J., & Niebur, E. (2002). Variable Resolution Displays: A Theoretical, Practical, and Behavioral Evaluation. *Human Factors.* (In Press.)

Pelz, J. B., Canosa, R., & Babcock, J. (2000). Extended Tasks Elicit Complex Eye Movement Patterns. In *Eye Tracking Research & Applications (ETRA) Symposium* (p. 37-43). Palm Beach Gardens, FL.

Pirenne, M. H. (1967). *Vision and The Eye* (2nd ed.). London, UK: Chapman & Hall.

Posner, M. I., Snyder, C. R. R., & Davidson, B. J. (1980). Attention and the Detection of Signals. *Experimental Psychology: General, 109*(2), 160-174.

Privitera, C. M., & Stark, L. W. (2000). Algorithms for Defining Visual Regions-of-Interest: Comparison with Eye Fixations. *IEEE Transactions on Pattern Analysis and Machine Intelligence (PAMI), 22*(9), 970-982.

Rayner, K. (1975). The Perceptual Span and Peripheral Cues in Reading. *Cognitive Psychology, 7,* 65-81.

Rayner, K. (Ed.). (1992). *Eye Movements and Visual Cognition: Scene Perception and Reading.* New York, NY: Springer-Verlag. (Springer Series in Neuropsychology)

Rayner, K. (1998). Eye Movements in Reading and Information Processing: 20 Years of Research. *Psychological Bulletin, 124*(3), 372-422.

Rayner, K., & Bertera, J. H. (1979). Reading Without a Fovea. *Science, 206,* 468-469.

Rayner, K., & Pollatsek, A. (1992). Eye Movements and Scene Perception. *Canadian Journal of Psychology, 46*(3), 342-376.

Rayner, K., Rotello, C. M., Stewart, A. J., Keir, J., & Duffy, S. A. (2001). Integrating Text and Pictorial Information: Eye Movements When Looking at Print Advertisements. *Journal of Experimental Psychology: Applied, 7*(3), 219-226.

Recarte, M. A., & Nunes, L. M. (2000). Effects of Verbal and Spatial-Imagery Tasks on Eye Fixations While Driving. *Journal of Experimental Psychology: Applied, 6*(1), 31-43.

Reddy, M. (1998). Specification and Evaluation of Level of Detail Selection Criteria. *Virtual Reality: Research, Development and Application, 3*(2), 132-143.

Reingold, E. M., Charness, N., Pomplun, M., & Stampe, D. M. (2002). Visual Span in Expert Chess Players: Evidence from Eye Movements. *Psychological Science*. ((In Press))

Reingold, E. M., Loschky, L. C., McConkie, G. W., & Stampe, D. M. (2002). Gaze-Contingent Multi-Resolutional Displays: An Integrative Review. *Human Factors*. (In Press.)

Robinson, D. A. (1968). The Oculomotor Control System: A Review. *Proceedings of the IEEE, 56*(6), 1032-1049.

Rosbergen, E., Wedel, M., & Pieters, R. (1990). *Analyzing Visual Attention to Repeated Print Advertising Using Scanpath Theory* (Tech. Rep.). University Library Groningen, SOM Research School. ((Technical Report 97B32))

Salvucci, D. D., & Anderson, J. R. (2001). Automated Eye-Movement Protocol Analysis. *Human-Computer Interaction, 16*, 39-86.

Salvucci, D. D., & Goldberg, J. H. (2000). Identifying Fixations and Saccades in Eye-Tracking Protocols. In *Eye Tracking Research & Applications (ETRA) Symposium* (p. 71-78). Palm Beach Gardens, FL: ACM.

Schmalstieg, D., & Schaufler, G. (1997). Smooth Levels of Detail. In *Proceedings of VRAIS'97* (p. 12-19). Albuquerque, NM.

Schoonard, J. W., Gould, J. D., & Miller, L. A. (1973). Studies of Visual Inspection. *Ergonomics, 16*(4), 365-379.

Schroeder, W. E. (1993a). Head-mounted computer interface based on eye tracking. In *Visual Communications and Image Processing'93 (VCIP)* (p. 1114-1124). Bellingham, WA.

Schroeder, W. E. (1993b). Replacing mouse and trackball with tracked line of gaze. In *Visual Communications and Image Processing'93 (VCIP)* (p. 1103-1113). Bellingham, WA.

Shebilske, W. L., & Fisher, D. F. (1983). Understanding Extended Discourse Through the Eyes: How and Why. In R. Groner, C. Menz, D. F. Fisher, & R. A. Monty (Eds.), *Eye Movements and Psychological Functions: International Views* (p. 303-314). Hillsdale, NJ: Lawrence Erlbaum Associates.

Sibert, L. E., & Jacob, R. J. (2000). Evaluation of Eye Gaze Interaction. In *Human Factors in Computing Systems: CHI 2000 Conference Proceedings*. ACM Press.

Smeets, J. B. J., Hayhoe, H. M., & Ballard, D. H. (1996). Goal-Directed Arm Movements Change Eye-Head Coordination. *Experimental Brain Research*, *109*, 434-440.

Snodderly, D. M., Kagan, I., & Gur, M. (2001). Selective Activation of Visual Cortex Neurons by Fixational Eye Movements: Implications for Neural Coding. *Visual Neuroscience*, *18*, 259-277.

Solso, R. L. (1999). *Cognition and the Visual Arts* (3rd ed.). Cambridge, MA: The MIT Press.

Sparks, D. L., & Mays, L. E. (1990). Signal Transformations Required for the Generation of Saccadic Eye Movements. *Annual Review of Neuroscience*, *13*, 309-336.

Starker, I., & Bolt, R. A. (1990). A Gaze-Responsive Self-Disclosing Display. In *Human Factors in Computing Systems: CHI '90 Conference Proceedings* (p. 3-9). ACM Press.

Stelmach, L. B., & Tam, W. J. (1994). Processing Image Sequences Based on Eye Movements. In *Conference on Human Vision, Visual Processing, and Digital Display V* (p. 90-98). San Jose, CA.

Tanriverdi, V., & Jacob, R. J. K. (2000). Interacting with Eye Movements in Virtual Environments. In *Human Factors in Computing Systems: CHI 2000 Conference Proceedings* (p. 265-272). ACM Press.

Todd, S., & Kramer, A. F. (1993). Attentional Guidance in Visual Attention. In *Proceedings of the Human Factors and Ergonomics Society, 37th Annual Meeting* (p. 1378-1382). Santa Monica, CA.

Tole, J. R., & Young, L. R. (1981). Digital Filters for Saccade and Fixation Detection. In D. F. Fisher, R. A. Monty, & J. W. Senders (Eds.), *Eye Movements: Cognition and Visual Perception* (p. 185-199). Hillsdale, NJ: Lawrence Erlbaum Associates.

Treisman, A. (1986). Features and Objects in Visual Processing. *Scientific American*, *255*(5), 114B-125,140.

Treisman, A., & Gelade, G. (1980). A Feature Integration Theory of Attention. *Cognitive Psychology*, *12*, 97-136.

Tsumura, N., Endo, C., Haneishi, H., & Miyake, Y. (1996). Image compression and decompression based on gazing area. In *Human Vision and Electronic Imaging*. Bellingham, WA.

Underwood, G. (Ed.). (1998). *Eye Guidance in Reading and Scene Perception*. Oxford, England: Elsevier Science Ltd.

Van Orden, K. F., & DiVita, J. (1993). Highlighting with Flicker. In *Proceedings of the Human Factors and Ergonomics Society, 37th Annual Meeting* (p. 1300-1304). Santa Monica, CA.

Velichkovsky, B., Pomplun, M., & Rieser, J. (1996). Attention and Communication: Eye-Movement-Based Research Paradigms. In W. H. Zangemeister, H. S. Stiehl, & C. Freksa (Eds.), *Visual Attention and Cognition* (p. 125-154). Amsterdam, Netherlands: Elsevier Science.

Vertegaal, R. (1999). The GAZE Groupware System: Mediating Joint Attention in Mutiparty Communication and Collaboration. In *Human Factors in Computing Systems: CHI '99 Conference Proceedings* (p. 294-301). ACM Press.

Vince, J. A. (1995). *Virtual Reality Systems.* Reading, MA: Addison-Wesley.

Von Helmholtz, H. (1925). *Handbuch der Physiologischen Optik (Treatise on Physiological Optics)* (Vol. III, Translated from the Third German ed.). Rochester, NY: The Optical Society of America.

Wallace, G. K. (1991). The JPEG Still Picture Compression Standard. *Communications of the ACM, 34*(4), 30-45.

Wang, M.-J. J., Lin, S.-C., & Drury, C. G. (1997). Training for strategy in visual search. *Industrial Ergonomics, 20,* 101-108.

Watson, B., Walker, N., & Hodges, L. F. (1997). Managing Level of Detail through Head-Tracked Peripheral Degradation: A Model and Resulting Design Principles. In *Virtual Reality Software & Technology: Proceedings of the VRST'97* (p. 59-63). ACM.

Watson, B., Walker, N., Hodges, L. F., & Worden, A. (1997). Managing Level of Detail through Peripheral Degradation: Effects on Search Performance with a Head-Mounted Display. *ACM Transactions on Computer-Human Interaction, 4*(4), 323-346.

Wedel, M., & Pieters, R. (2000). Eye Fixations on Advertisements and Memory for Brands: A Model and Findings. *Marketing Science, 19*(4), 297-312.

Wolfe, J. M. (1993). Guided Search 2.0: The Upgrade. In *Proceedings of the Human Factors and Ergonomics Society, 37th Annual Meeting* (p. 1295-1299). Santa Monica, CA.

Wolfe, J. M. (1994). Visual Search in Continuous, Naturalistic Stimuli. *Vision Research, 34*(9), 1187-1195.

Wolfe, J. M., & Gancarz, G. (1996). GUIDED SEARCH 3.0: A Model of Visual Search Catches Up With Jay Enoch 40 Years Later. In V. Lakshminarayanan (Ed.), *Basic and Clinical Applications of Vision Science* (p. 189-192). Dordrecht, Netherlands: Kluwer Academic.

Wooding, D. (2002). Fixation Maps: Quantifying Eye-Movement Traces. In *Eye Tracking Research & Applications (ETRA) Symposium.* New Orleans, LA.

Yarbus, A. L. (1967). *Eye Movements and Vision.* New York, NY: Plenum Press.

Young, L. R., & Sheena, D. (1975). Survey of Eye Movement Recording Methods. *Behavior Research Methods & Instrumentation, 7*(5), 397-439.

Zee, D. S., Optican, L. M., Cook, J. D., Robinson, D. A., & Engel, W. K. (1976). Slow Saccades in Spinocerebellar Degeneration. *Archives of Neurology, 33,* 243-251.

Zeki, S. (1993). *A Vision of the Brain.* Osney Mead, Oxford: Blackwell Scientific Publications.

Zhai, S., Morimoto, C., & Ihde, S. (1999). Manual and Gaze Input Cascaded (MAGIC) Pointing. In *Human Factors in Computing Systems: CHI '99 Conference Proceedings* (p. 246-353). ACM Press.

Zorin, D., & Schröder, P. (2000). *Course 23: Subdivision for Modeling and Animation.* New York, NY. (SIGGRAPH 2000 Course Notes, URL: <http://www.mrl.nyu.edu/dzorin/sig00course/> (last accessed 12/30/00))

Zorin, D., Schröder, P., & Sweldens, W. (1997). Interactive Multiresolution Mesh Editing. In *Computer Graphics (SIGGRAPH '97).* New York, NY.

Index

aberration
- chromatic, 20
- spherical, 20

Adaptive Control of Thought-
 Rational (ACT-R), 210

adaptive thresholding, 126

advertising, 193
- "wearout", 196
- headline, 197
- packshot, 197

algorithmic Regions Of Interest
 (aROIs), 161

Allopenna, Paul D., 168

α cells, 25, 26

Anders, Geerd, 171

Andre, Anthony D., 174

apparent motion, 38

apparent movement, 35

Application Program Interface (API),
 95

Applied Science Laboratories (ASL),
 168

Arbib, Michael A., 29

Area V1, 18

Area V2, 18

Area V3, 18

Asaad, Wael F., 137

attention
- spotlight of, 9

attentional
- glue, 11
- window, 11

attentional feedback loop, 19

attenuation filter, 7

Ballard, Dana H., 165, 166, 178

Behavior Research Methods, Instruments, and Computers (BRMIC),
 75

Bertera, James H., 141, 157

β cells, 25, 26

Bhagwat, Sameer, 187

bipolar cells, 22

blinks, 113

Bolt, Richard, 206

bottom-up model, 12

boundary paradigm, 141

Broadbent, Donald, 6

Buswell, G. T., 148

Caird, Jeff K., 177

Campfire on Perceptually Adaptive
 Graphics, 225

cells
- α, 25, 26
- β, 25, 26
- amacrine, 22
- bipolar, 22, 23
- cortical
-- complex, 27
-- simple, 27
- ganglion, 22, 24, 27
- horizontal, 22
- inner nuclei, 22
- M, 25